CLAIMS AGAINST IRAQI OIL AND GAS

Legal Considerations and Lessons Learned

This book presents the first and only comprehensive examination of the legal issues surrounding international debt recovery on claims against Iraqi oil and gas. It offers a snapshot view of Iraq's outstanding debt obligations and an analysis of the significance of the theory of odious debt in the context of the Iraqi situation. The list of legal issues examined includes relevant provisions of the 2005 Iraqi Constitution, controlling United Nations Security Council resolutions, pertinent articles of the 2007 Kurdistan Regional Government oil and gas law (No. 22) and the many nuanced and technical questions raised thereby, legal pronouncements of Iraq and selected other nations that aim to protect Iraqi oil and gas, and general problems associated with recognition and enforcement of awards or judgments that may involve such oil and gas or revenues from their sale. Also discussed are the lessons learned by the handling of the Iraqi debt experience and the transferability of those lessons to future situations in which resource-rich nations may have outstanding financial obligations to other members of the world community or their nationals.

Rex J. Zedalis is a member of the faculty at the University of Tulsa, College of Law, where he has received numerous awards for his excellent teaching, including an award in 2004 for the university's most outstanding professor. He also has been recognized for his extensive publication record in both U.S. and European international law journals. He is currently the Phyllis Hurley Frey Professor of Law and the Director of the Comparative and International Law Center (CILC) at the University of Tulsa. He has served as a Fellow with, and a former Assistant Director of, the College of Law's National Energy Law & Policy Institute. Professor Zedalis also acted as consultant to international organizations, foreign governments, and domestic as well as international law firms during his thirty-five-year career. Professor Zedalis is the author of *The Legal Dimensions of Oil and Gas in Iraq: Current Reality and Future Prospects*.

Claims Against Iraqi Oil and Gas

LEGAL CONSIDERATIONS AND LESSONS LEARNED

Rex J. Zedalis

University of Tulsa

CAMBRIDGE
UNIVERSITY PRESS

CAMBRIDGE UNIVERSITY PRESS
Cambridge, New York, Melbourne, Madrid, Cape Town, Singapore,
São Paulo, Delhi, Dubai, Tokyo, Mexico City

Cambridge University Press
32 Avenue of the Americas, New York, NY 10013-2473, USA

www.cambridge.org
Information on this title: www.cambridge.org/9780521193504

First published 2010

Printed in the United States of America

A catalog record for this publication is available from the British Library.

Library of Congress Cataloging in Publication data

Zedalis, Rex J.
Claims against Iraqi oil and gas : legal considerations and lessons learned / Rex J. Zedalis.
p. cm.
Includes bibliographical references and index.
ISBN 978-0-521-19350-4 (hardback)
1. Debts, External – Law and legislation – Iraq. 2. Petroleum law and legislation – Iraq.
3. Natural gas – Law and legislation – Iraq. 4. Oil and gas leases – Iraq.
5. Revenue sharing – Law and legislation – Iraq. 6. Petroleum industry and trade – Iraq.
I. Title.
KMJ3532.Z43 2010
343.567′0772 – dc22 2009051478

ISBN 978-0-521-19350-4 Hardback

This book is dedicated to my wife, Catherine, and my two sons, Ian and Bryce, who sustain, humor, and endure me through thick and thin, as well as to my many teachers, mentors, and colleagues who have served over the years as models of the academic scholar. Special recognition is deserved by David S. Clark (Willamette University), Walter Gellhorn (Columbia University), W. T. Mallison, Jr. (George Washington University), and Oscar Schachter (Columbia University). The support of the University of Tulsa, College of Law, Faculty Research Fund; the funding associated with holding the Phyllis Hurley Frey Professorship; and the encouragement and enthusiasm of the College of Law's Dean, Janet Koven Levit, are deserving of particular acknowledgment.

CONTENTS

PREFACE

The general topic of legal claims against any foreign nation for debts owed to other nations raises an exceptionally long and involved list of intricate and complex issues from susceptibility to lawsuits, to jurisdiction, to potential defenses, to enforcement of judgments, and numerous associated matters. Similarly, the topic cannot avoid the possible significance related to distinguishing between claimants that are governments and those that are private individuals or legal entities, as well as the practical ramifications that emerge from the existence and location of assets from which any successfully sought judgment may be satisfied. Similarly, there are the questions of whether the relief desired is to proceed from an adjudicative- or arbitral-type process and whether such a process is to (or must) take place before an international or national tribunal.

Broadly speaking, whenever one foreign sovereign owes a form of debt to another sovereign, the traditional mechanism of diplomacy is most often employed to secure payment. When not completely efficacious, the creditor nation may resort to alternatives to seek satisfaction, including various types of internationally mediated, arbitrated, or adjudicated claim-resolution processes. Clearly, a substantial portion of the debts claimed by one foreign sovereign against another involve debts with their fundamental origins in obligations owed by the creditor sovereign to private individuals or private legal entities considered nationals of the foreign sovereign advancing the claim. This is because most often, the rules of international law are viewed as speaking not to nationals but rather to the nation-states that protect and represent them.

When causes of action have been maintained in national fora, whether adjudicative or otherwise, it is more often the result of the

initiative of private individuals or private legal entities. Although at the level of international claims resolution it is typically the government of the claimant that advances the claim, before any national fora, the claimant is recognized as vested with the authority to speak for itself. In some instances, claims are initiated before national fora because of an absence of an international mechanism established to address the claims of natural or juridical persons; in other cases, the claims may proceed from dissatisfaction with the results from the use of any such mechanism and an attempt to sidestep compulsory resort thereto. Regardless of the reason, actions brought in such fora can present many of the issues raised herein. The nation of the forum in which the action is commenced may endorse certain notions about the limited susceptibility of foreign sovereign creditors to prosecution. The party maintaining the action may face hurdles obtaining effective jurisdiction allowing it to move forward or find the creditor nation without sufficient assets to permit full satisfaction from any judgment successfully received.

It is well known that Iraq's substantial oil and gas reserves and the potential for huge amounts of revenues from international crude oil exports place it in a position of having assets that can be applied to extinguishing debts owed to other nations, private individuals, and private legal entities. Estimates suggest that proven Iraqi crude-oil reserves are approximately 115 billion barrels,[1] with unproven reserves totaling perhaps more than 100 billion barrels.[2] Through 2008 and into early 2009, Iraq was producing only 2.4 million to 2.5 million barrels per day (mbpd),[3] exporting approximately 2 mbpd[4]; however, before the two Gulf Wars, it had reached production levels of 3.5 mbpd[5] and has its

[1] See Energy Information Administration, U.S. Dep't of Energy, Official Energy Statistics from the U.S. Government (Aug. 2007), available at www.eia.doe.gov/emeu/cabs/Iraq/Oil.html (accessed Jan. 14, 2009).

[2] See id. See also Weshad Rassak, "Iraq: Private Ownership of Oil and the Quest for Democracy," MPRA Paper No. 54 (Oct. 10, 2007) (100 billion barrels); "Foreign Firms Covet Iraq Oil Reserves, Global Energy," *Iran Daily* (Feb. 6, 2006) (265 billion barrels), available at www.iran-daily.com/1384/2494/html/energy.htm (accessed Jan. 14, 2009).

[3] See Bureau of Near Eastern Affairs, U.S. Dep't of State, Iraq Weekly Status Report, Dec. 31, 2008, at 18 (graph), available at www.state.gov/documents/organization/113677.pdf (accessed Jan. 14, 2009).

[4] This is based on the fact that Iraqi domestic consumption has run around 500,000 bpd.

[5] See U.S. Energy Information Administration, U.S. Dep't of Energy (1998), available at www.arabchamber.com/arab-countries/iraq/Y/iea2.htm (see graph) (accessed June 19, 2008).

eye on a production level of 6 mbpd,[6] with most supplied to the international market. Even at current production and export levels, when international crude prices return to prerecession heights, Iraq could have revenues that exceed by tens of millions of dollars its per-year budget expenditures. To illustrate this point, during the height of summer 2008 international crude-oil price run-up, indications were that Iraqi oil-export revenues were on track to outstrip budget expenditures by almost $80 billion.[7]

Recognizing the potential for Iraq to use oil revenues to satisfy outstanding debts, as early as Spring 1991 the UN Security Council established in Security Council resolution 687 the UN Compensation Commission (UNCC) to oversee and address claims against Iraq for debts reportedly incurred because of the invasion of Kuwait by the military forces of Saddam Hussein in August 1990 – the invasion that led to the First Gulf War.[8] Oil revenues were to be used to satisfy relevant debts entitled to payment under the UNCC rules. Notably, debts antedating the invasion or not resulting from the military action and the consequent wartime civil disorder or hostage taking were not cognizable under the UNCC mandate. In other words, the UNCC was not designed to consider all debts claimed to be owed by Iraq, regardless of when they were accrued or the circumstances under which they emerged. Iraqi oil revenues were to serve as the basis for compensating claimants against that nation but only with respect to debts said to be directly attributable to the military action involving the invasion of Kuwait.

Many other claimants against Iraq remained on the sidelines – including those with debts predating the Kuwait military operations and the First Gulf War and debts not arising from military action, civil disorder, or hostage taking, as well as those with claims that developed as a result of post–First Gulf War relations or dealings with Iraq. It also was believed to be important to ensure that Iraqi oil revenues

[6] See Ahmed Rasheed, "Iraq Opens 11 Oil and Gas Fields for Foreign Bids," *AFX News* (Dec. 31, 2008), available at www.freerepublic.com/focus/f-news/2156746/posts (accessed Jan. 14, 2009).

[7] See James Glanz & Campbell Robertson, "As Iraq Surplus Rises, Little Goes Into Rebuilding," *New York Times* (Aug. 5, 2008), available at www.nytimes.com/2008/08/06/world/middleeast/06surplus.html (accessed Jan. 14, 2009).

[8] See UN Security Council Res. 687, UN SCOR, 46th Sess., 2981st mtg., UN Doc. S/RES/687 (1991), especially at paras. 16–19.

would be preserved from creditor claims in order to be available for any future rebuilding of that country. Perhaps in partial acknowledgment of these facts, the UN Security Council adopted a series of resolutions insulating Iraqi hydrocarbon revenues, as well as the hydrocarbons themselves, from legal action, directing that member states adopt their own domestic measures to implement that UN obligation.[9] Additionally, the Iraqi government, including the Kurdistan Regional Government (KRG), in order to provide adequate flexibility regarding oil and gas contracts and other commitments earlier entered into, inserted provisions in controlling legal enactments addressing reconsideration and alteration of such contracts and commitments, the effect being to limit the legal consequences for moving to modify or change previous arrangements with international oil and gas companies. The significance of both actions for those owed debts by or in contracts with Iraq cannot be understated, especially in view of the fact that 95 percent of the country's revenues come from the sale and export of crude oil and oil products, which any effective legal recourse against Iraq would be compelled to have as the object of its remedial desires.[10] Stated another way, the insulation by the Security Council of Iraqi hydrocarbons and proceeds of their sales – as well as the legal protection under Iraqi law for reconsideration of certain contractual commitments – would appear to seriously complicate the ability of claimants to secure satisfaction for Iraqi failures to completely fulfill obligations owed to the claimants.

In view of the foregoing discussion, the central and exclusive focus of this book is on the various legal issues and juridical nuances related to two matters: (1) the specific international measures (from the Security Council and other bilateral agreements) and associated domestic enactments designed to protect Iraqi hydrocarbons and revenues

[9] See UN Security Council resolution 986 (1995) at paras. 14–15, available at www.iraqwatch.org/un/unscresolutions/s-res-986.htm (accessed Jan. 14, 2009); resolution 1483 (2003) at paras. 22–23, available at www.iraqwatch.org/un/unscresolutions/s-res-1483-052203.pdf (accessed Jan. 15, 2009); resolution 1546 (2004) at para. 27, available at www.iraqwatch.org/un/unscresolutions/s-res-1546.pdf (accessed Jan. 15, 2009); and resolution 1790 (2007) at para. 3, available at http://daccessdds.un.org/doc/UNDOC/GEN/N07/650/72/PDF/N0765072.pdf?OpenElement (accessed Jan. 15, 2009).

[10] See "Blood and Oil: How the West Will Profit from Iraq's Most Precious Commodity," *The Independent* (7 Jan. 2007), available at www.independent.co.uk/news/world/middle-east/blood-and-oil-how-the-west-will-profit-from-iraq's-most-precious-commodity-431119.html (accessed Jan. 15, 2009).

generated from being available to meet creditor claims; and (2) those particular provisions of Iraqi national law – constitutional and statutory – that appear designed to provide protection from legal claims resulting from alleged alterations by Iraq in the terms of contractual commitments concerning oil and gas. Most of this book's attention is devoted to examining the meaning and significance of the relevant Security Council formulations associated with the first item, and the legal measures implemented in the United States to execute those formulations, as well as the second item dealing with relevant provisions of Iraqi law addressing changes in contractual rights and duties. Nevertheless, regarding the protection of hydrocarbons and revenues from claims by creditors, attention also is given to measures adopted in selected other countries in which potential claimants are likely to reside or Iraqi assets are likely to be found. Additional discussion pertains to certain relevant parallel provisions of the U.S.–Iraq Status of Forces Agreement (SOFA) that primarily governs the maintenance of foreign forces in Iraq for the next few months to several years. As shown in the particular discussions, the words of all relevant provisions are parsed closely, with specific terms the focus of one chapter and other terms from the same provisions the focus of subsequent chapters.

With regard to UN Security Council formulations, paragraph 22 of resolution 1483[11] – extended first until the end of 2009 by the terms of resolution 1859, which was adopted in the waning days of 2008, and then until the end of 2010 by resolution 1905, adopted December 21, 2009 – it provides that "petroleum, petroleum products, and natural gas originating in Iraq shall be immune, until title passes to the initial purchaser from legal proceedings" and "proceeds and obligations arising from sales [of such] shall enjoy privileges and immunities equivalent to those enjoyed by the United Nations."[12] Article 26 of the SOFA between the United States and Iraq, adopted just before resolution 1859, recognizes the need "to continue to safeguard Iraq's revenues from oil and gas" and the fact that the debts of the former Iraqi regime caused the "President of the United States . . . to exercise[] his authority to protect from United States judicial process the Development Fund

[11] See resolution 1483, supra note 9 at para. 22.

[12] See Security Council resolution 1859 (22 Dec. 2008) at para. 1, available at http://daccessdds.un.org/doc/UNDOC/GEN/N08?666?05/PDF/N0866605.pdf?OpenElement (accessed Jan. 27, 2009).

for Iraq and certain other property in which Iraq has an interest."[13] With regard to pertinent constitutional and statutory measures of Iraqi national law, article 141 of the Iraqi Constitution indicates that legislation of the KRG since 1992, as well as decisions of the government including court decisions and "contracts, shall be considered valid unless they are amended or annulled" under laws of the KRG and so as not to contradict the Iraqi Constitution.[14] Article 54 of the 2007 KRG oil and gas law (No. 22) provides that "agreements related to Production Sharing Contracts entered into by the [KRG] prior to the entry into force of this Law" must be reviewed for consistency therewith and decisions regarding them "shall be final." The article also provides that "authorizations and memoranda of understanding related to oil and gas . . . signed by the [KRG] prior to entry into force of this Law" shall be "null and void" unless approved within the KRG.[15] The simple clarity of these provisions obscures the many complexities related to their precise wording – complexities that raise genuine questions about the provisions' ability to accomplish the "short-circuiting" role they are all designed to play. The nature of the Iraqi assets protected, the obligations they serve to financially secure, the point at which those assets become fair game, the relationship between article 141 of the Constitution and article 54 of the KRG oil and gas law, the existence of other constitutional provisions limiting contract amendment or annulment, and the significance of an agreement related to a production sharing contract (PSC) as opposed to an authorization or Memorandum of Understanding (MOU) all merit sedulous and close examination.

To set the general problem of debts owed by Iraq in the context of the real world, it must be noted that various legal claims have been lodged and adjudicated against Iraq in American courts since the period after that country's invasion of Kuwait that led to the First Gulf War. Illustrative cases include *Consarc Corp. v. Iraqi Ministry of*

13 See U.S.–Iraq Status of Forces Agreement, art. 26, Iraq Oil Report, Breaking: Text of Status of Forces Agreement (Nov. 18, 2008), available at www.iraqoilreport.com/2008/11/18/breaking-text-of-status-of-forces-agreement (accessed Nov. 28, 2008).

14 See Iraqi Constitution, art. 141, available at www.export.gov/iraq/pdf/iraqi_constitution.pdf (accessed Nov. 15, 2007). Also available at www.uniraq.org/documents/iraqi_constitution.pdf (accessed Mar. 5, 2009).

15 See Oil and Gas Law (No. 22) of the Kurdistan Region–Iraq (2007), art. 54, available at www.krg.org/uploads/documents/Kurdistan%20Oil%20and%20Gas%20Law%20English_2007_09_06_h14m0s42.pdf (accessed Jan. 27, 2009).

Industry & Minerals, *Brown & Root International, Inc. v. State Company for Oil Projects*, and *Commercial Bank of Kuwait v. Rafidain Bank*, all involving claims of debts owed by Iraq and arising from circumstances prior to the First Gulf War. In view of the inability of claimants to seek recovery under the terms of the very narrow and specific UNCC mandate, and the absence of any other mechanism established by agreement between the United States and Iraq for resolving the claims, petitioners were left with no option other than to pursue relief in an appropriate national court.

In *Consarc*,[16] the New Jersey engineering and production firm had contracted in 1989 with the Iraqi Ministry of Industry & Minerals to supply four powerful laser furnaces associated with the manufacture of specialty metals. Iraq declared to the firm and the U.S. government export-licensing agency that the furnaces were to be used in the production of prosthetics. Subsequently, intelligence reports indicated that the furnaces were actually to be used in Iraq's nuclear-weapons program, and the appropriate export licenses were thus withdrawn. Consarc contended that the actions of Iraq leading to the blocking of the export constituted fraud and a breach of contract for which compensation was payable. The August 1991 invasion of Kuwait by Iraqi forces resulted in the freezing of all Iraqi funds and assets subject to U.S. jurisdiction, thereby compelling Consarc and others similarly situated to secure authorization of the U.S. Department of Treasury's Office of Foreign Assets Control (OFAC) prior to instituting legal action against Iraq. Consarc sought and received such authorization and commenced litigation with a September 1990 filing in the U.S. District Court in Washington, DC. After an April 1991 hearing, Consarc received a favorable summary judgment, which included a $6.4 million compensatory and a $55 million punitive-damage award.[17] In the estimation of the Court, Consarc and its U.K. subsidiary had "fully performed all their obligations . . . [and the Ministry of Industry & Minerals]

[16] *Consarc Corp. v. Iraqi Ministry of Industry & Minerals*, 1991 U.S. Dist. LEXIS 4884 (D. D.C. Apr. 10, 1991).

[17] Several follow-on appeals, litigating various related issues, resulted from the District Court's decision. See, generally, *Consarc Corp. v. Iraqi Ministry*, 27 F. 3d 695 (D.C. Cir. 1994), *Consarc Corp. v. U.S. Treasury Dep't Office of Foreign Assets Control*, 871 F. Supp. 1463 (D. D.C. 1994), and *Consarc Corp. v. U.S. Treasury Dep't Office of Foreign Assets Control*, No. 94–5390 (decided Dec. 15, 1995) (D.C. Cir. 1995), available at www.ll.georgetown.edu/federal/judicial/dc/opinions/94opinions/94-5390a.html (accessed Jan. 16, 2009).

had . . . committed egregious and wantonly malicious and willful acts of fraud and breaches of contract . . . including the knowing submission of a fraudulent end-user certificate [required for exportation]."[18]

In *Brown & Root International*,[19] the predecessor of what is known now as Kellogg, Brown & Root (KBR) had performed extensive repair work on Iraq's offshore oil terminals in the Persian Gulf at Mina al-Bakr, in the far south of the country near the border with Kuwait. According to the plaintiff, the services performed were valued at nearly $18 million and it sought a default judgment for Iraq's failure to make payment. In June 1992, Brown & Root was granted a judgment that compensated it for its unpaid services.

Commercial Bank[20] involved another instance of legal action against an organ or appendage of the Iraqi government, this time for failure to repay certain loans secured by letters of credit and obligations taken on by Rafidain Bank, a wholly owned entity of the Iraqi government, and guarantees of Rafidain's repayment made by the Central Bank of Iraq (CBI; the equivalent of the U.S. Federal Reserve). The loans made to Iraq and secured by the letters of credit, obligations, and guarantees were all issued from Commercial Bank of Kuwait and a syndicate of other banks in the sum total of $1.1 billion. Commercial Bank's share of that total amounted to $33 million, of which it claimed Rafidain Bank and the CBI had defaulted in repaying $7.4 million after the invasion of Kuwait and subsequent outbreak of the First Gulf War. The case was referred by the U.S. Federal Court for the Southern District of New York to a magistrate for full consideration of the issues presented. After the magistrate's report, the Court in late 1992 granted judgment on the CBI's motion for a default judgment, awarding $71 million in damages.

It is clear that the nature of legal issues confronted by U.S. courts in *Consarc*, *Brown & Root International*, and *Commercial Bank of Kuwait* involved matters outside the scope of this book. Rather, the focus is limited to questions arising from complexities of the international and national measures specifically insulating Iraqi oil and gas – and the revenues generated – from creditor claims, as well as questions arising

[18] See *Consarc*, supra note 16 at 3.

[19] See *Brown & Root International, Inc. v. State Company for Oil Projects*, 91 Civ. 594 (S.D. Tex. 1991).

[20] See *The Commercial Bank of Kuwait v. Rafidain Bank*, 91 Civ. 6500 (S.D.N.Y. Oct. 1, 1992). The District Court case is also described in *Commercial Bank of Kuwait v. Rafidain Bank*, 15 F.3d 238 (2d Cir. 1994).

from the comparable complexities of the terms of various provisions of Iraqi law addressing Iraqi changes in contractual obligations, including those dealing with oil and gas. Although it is accurate to assume that many of the matters wrestled with in the court decisions referenced can be presented in the context of legal disputes involving the matters focused on herein, readers should not expect to gain any insight about such in the chapters that follow. If it does not concern the legal twists and turns of the international or national legal measures insulating Iraqi hydrocarbons and the proceeds of their sale from claims by those contending that Iraq owes some debt, or the implications of those Iraqi legal measures concerning alterations in Iraq's contractual commitments, then it is not the concern of this book. Issues about the efficaciousness of OFAC's power and authority to block lawsuits against foreign assets that it had previously frozen, or the terms and provisions of legislative measures that protect foreign sovereigns against action in national courts, are important in their own right. However, the purpose of the current book is limited and centers on two specific and significant matters: (1) the international and domestic measures that insulate Iraqi oil and gas and proceeds from their sale from legal action; and (2) the Iraqi measures designed to prevent legal action as a result of Iraq altering its contractual obligations with others.

Two final yet extremely important observations are warranted prior to venturing into the specifics of this book. The first addresses the fact that article 25 of the UN Charter clearly and explicitly recognizes resolutions of the Security Council that represent decisions of that body as binding on all member states.[21] In view of the terms and provisions of international agreements constituting one of the long-acknowledged sources of international law,[22] the effect of article 25 is to make the obligations of Security Council resolutions 1483, 1859, and 1905 – referenced previously[23] and in the chapters that follow – part and parcel of the corpus of international law from which the community of nations cannot escape. The second observation concerns where the insulation

[21] See UN Charter, art. 25, 59 Stat. 1031, T. S. No. 993, 3 Bevans 1153, 24 U.S.T. 2225, T.I.A.S. No. 7739, available at www.un.org/en/documents/charter/ (accessed Sept. 19, 2009).

[22] See Statute of the International Court of Justice, 59 Stat. 1055, T.S. No. 993, 3 Bevans 1179, available at www.icj-cij.org/documents/index.php?p1=4&p2=2&p3=0 (accessed Sept. 19, 2009).

[23] See text accompanying supra notes 11–13.

from creditor claims provided to Iraqi oil and gas by those Security Council resolutions and other provisions of law is situated within the larger context of general international law. This not only implicates the first observation but also calls for recollection of the basic notion of the general legal status of a sovereign and people engaged in war who then regain governmental control over territory earlier lost as a consequence of military operations or subsequent foreign occupation. In the Iraqi situation – in which contractual obligations and debts were incurred by that nation before the outbreak of the First Gulf War in 1991, in the interwar years preceding the Second Gulf War in 2003, and even in the subsequent period when under the control of the U.S.-led coalition occupying forces – this requires reference to what international law states about (1) the effect of war, hostile military invasion, and possible foreign occupation on the dealings of a sovereign whose self-rule is later restored; and (2) the situation's effect on contracts and obligations that may have been transacted by the legitimate sovereign under those or similar circumstances.

Regarding the former – the general effect of war, invasion, and occupation on actions and authority of a later-returned sovereign – a foundational principle of international law and most legal systems is that of *ex injuria jus non oritur*, loosely translated as "illegal acts cannot create law" or "a right cannot arise from a wrong."[24] Although the twists and turns of the principle can be many, at root and as a sweeping generalization, it has the effect of preventing military activities from altering the legal status quo ante, at least to the extent that the territory affected by such activities returned to the control of the sovereign who exercised authority prior to the military operations. Apparently dating from Roman times[25] and thoroughly explored by Hugo Grotius[26]

[24] See, generally, Brigette Stern, "Dissolution, Continuation and Succession in Eastern Europe" (1998); and Gerard Kreijen, "State Failure, Sovereignty and Effectiveness" (2004), the latter contrasting the principle with that of *ex factis jus oritur*, translated as "the facts create the law."

[25] For a discussion of the ancient doctrine, see Randall Lesaffer, "Gentili's *ius post bellim* and early-modern peace treaties" 32–35, available at www.law.nyu.edu/idcplg?IdcService=GET_FILE&dDocName=ECM_DLV_016258&RevisionSelectionMethod=LatestRevision (accessed Sept. 19, 2009).

[26] See Hugo Grotius, *De Jure Belli ac Pacis*, Bk. III, Chpt. 9 (1625), available at www.constitution.org/gro/djbp_309.htm (accessed Sept. 19, 2009). See also Emmerich de Vattel, "The Law of Nations or the Principles of Natural Law," Bk. 3, Chpt. 14, Secs. 204–222 (1758), available at www.lonang.com/exlibris/vattel/vatt-314.htm (accessed Sept. 19, 2009).

(a founding father of international law), the doctrine of *postliminium* proved a natural outgrowth of the *ex injuria* principle. The thrust of the doctrine of *postliminium*, although rife with many twists and turns of its own, proclaims the situation and status of the original sovereign and its dominated people to remain unaltered by hostile military presence or occupation, once the legitimate sovereign and its people are returned to power.[27] Today, the doctrine can be seen in part as a historical antecedent, given that much of what the doctrine sought has been codified in the laws of war – specifically, the Regulations Respecting the Laws and Customs of War on Land, annexed to Hague Convention (IV) of 1907,[28] and the various 1949 Geneva Conventions.[29] Regarding the effect of the displaced sovereign's laws and property interests associated with the territory, articles 43, 46, 53, and 55 of the Regulations Respecting the Laws and Customs of War on Land seem especially significant. Article 43 indicates that the displaced sovereign's laws remain in force unless absolutely prevented by military necessity, whereas article 46 protects private property from confiscation and articles 53 and 55 regard state-owned property as available for the hostile power's use only pursuant to particular standards.

Regarding what international law states about war, invasion, and occupation's effect on contracts and other obligations transacted by a legitimate sovereign whose self-rule is eventually restored – the latter of the concepts referenced previously – it seems that a body of law more national than international in character and whose thrust is complementary to that of *ex injuria*, *postliminium*, and the laws of war has developed within the community of states. This law of international transactions accepts the fact that war and activities associated with military hostilities can serve to act on contracts and other obligations in a variety of different ways. First, they may render performance illegal, as in the case of contracts and obligations between enemy states.[30] Second, they may provide one party with the defense of frustration for

27 See, generally, *Hall's International Law* at 577–585 (ed. Pearce Higgins, 8th ed. 1924).

28 Available at www.icrc.org/IHL.NSF/INTRO/195?OpenDocument (accessed Sept. 19, 2009).

29 See, e.g., Convention Relative to the Protection of Civilian Persons in Time of War (1949), 6 U.N.T.S. 3516, T.I.A.S. No. 3365, 75 U.N.T.S. 287.

30 See Samuel Williston, "The Law of Contracts" Sec. 1747 (1922); *Zinc Corp. v. Hirsch* [1916] 1 K.B. 541 (1922).

its failure to perform its commitments.[31] Third, they may present the problem of performance – interfered with by hostilities – that can be resumed or completed following cessation of those hostilities.[32] Fourth, they may raise the distinction between contracts and obligations not yet executed and those that have been completed and involve the accrual of a right of action for a debt owed and not paid.[33] Long ago, it was observed that no general principle of law provides for the abrogation of contracts as a result of war.[34] Nevertheless, some authority exists for viewing war as ending even those contracts whose performance could be reinstated following the conclusion of hostilities,[35] although support for the contrary position also must be admitted.[36] Concerning situations in which performance has been completed and there remains only payment to be made at the time of the outbreak of hostilities, at least Anglo-American jurisprudence suggests that war has no effect on the continuation of the debt owed.[37] Indeed, that is an underlying assumption inherent in at least a 1994 Second Circuit decision[38] following up on *Commercial Bank of Kuwait*, examined previously in conjunction with *Consarc* as well as *Brown & Root International.* After all, in the Second Circuit's consideration in *Commercial Bank*, the Iraqi authorities even raised the First Gulf War by way of an important circumstance affecting their ultimate liability. Neither they nor the court, however, came close to suggesting that war somehow exculpated them from having to pay the debt they owed.[39] Furthermore, the notion that accrued debts are to be paid – despite the fact that

[31] See Arnold D. McNair, "Legal Effects of War," Chpt. 6, "Frustration of Contract" (1944).

[32] See id. at 83, acknowledging such but indicating that it is a rare event.

[33] See id. at 101–107.

[34] See, e.g., *Ertel Beiber & Co. v. Rio Tinto Co.* [1918] A.C. 260, 269; *Ottoman Bank v. Jebara* [1928] A.C. 269, 276.

[35] See Arnold D. McNair, supra note 30 at 98–99.

[36] See, e.g., Amos S. Hershey, "The Essentials of International Public Law," 368–369 (1918).

[37] See, e.g., *Shering Ltd. v. Stockholms Enskilda Bank Aktiesbolag and Others*, 12 Int'l Law Rpts. [1943–45] 308 (1951). See also "Effects of War on Contract," 31 *Harvard L. Rev.* 640–643 (1917–18).

[38] See *Commercial Bank of Kuwait v. Rafidain Bank*, 15 F.3d 238 (2d Cir. 1994), available at http://bulk.resource.org/courts.gov/c/F3/15/15.F3d.238.93-7552.533.html (accessed Sept. 20, 2009).

[39] See *Commercial Bank of Kuwait*, id. (paras. 34–35 of the opinion discussing the inability of the Gulf War to affect a finding of "willfulness" of Iraqi default).

military operations may have intervened between the debtor and creditor states – suggests a consonance of this field of law with that involving the general international law concepts of *ex injuria*, *postliminium*, and the laws of war.

Reviewing the previous discussion in its totality, what is revealed is an international-law approach that essentially preferences the resumption of the same situation – presumably including the contract debt situation – that prevailed before military operations and any subsequent foreign occupation, at least after the legitimate sovereign has been restored to power. Situating Iraqi debts in that type of international-law context certainly favors Iraq's creditors, except for the fact that the international community acted through the Security Council and other mechanisms to provide Iraqi oil and gas and revenues from their sale with a certain level of protection from those creditors. In other words, whereas the community of nations may have participated over many centuries in the development of both international and national principles, doctrines, and codified provisions of law that seem to favor the resumption of the preexisting situation following the ending of hostilities and the return of the earlier sovereign, those same nations are similarly empowered to take the prevailing law in another and entirely different direction if they so choose. Actions of the UN such as those reflected in Security Council resolutions 1483, 1859, and 1905, and actions of various individual nation-states seem to represent efforts to push the relevant international and national law in a direction beneficial to Iraq and clearly distinct from what otherwise appears to be the case. In the chapters that follow, the complexities, nuances, and intricacies of the various legal measures employed to accomplish that task serve as the central and exclusive focus of this book's attention.

Tulsa, Oklahoma
January 2010

Part One

The existence of Iraqi debt to foreign creditors, both governmental and private, seems beyond question; few developing countries, although blessed with substantial natural resources, could escape some degree of financial dependence on the developed countries of the world community. Apparently, the only debatable question in Iraq's case concerns the level of debt owed to foreign creditors. A purpose of Part One of this book on the legal considerations related to causes of action against Iraqi oil and oil revenues for debts owed by Iraq is to distill the information currently in the public domain regarding the extent of that nation's debt. Despite the fact that those estimates vary widely, it is clear that in all instances they fix total debt figures that are quite substantial, especially given the current economic needs faced by the Iraqi government. These needs reflect not only the years of infrastructure neglect witnessed during the Saddam Hussein regime but also the ravages of the Second Gulf War, as well as the civil unrest and ethnic and sectarian rivalry that followed it. The situation is further complicated by the fact that revenues resulting from oil and gas exports have yet to generate sufficient surplus national income to allow meaningful satisfaction of what is owed to creditors.

Part One also examines the early raised contention that much of what was owed by Iraq as a consequence of negotiations and dealings successfully undertaken during Saddam Hussein's regime with foreign governments and private parties should be characterized as "odious debt." Therefore, this debt cannot continue as an obligation of subsequent replacement Iraqi governments. At least, the notion of odious debt has been controversial in international legal circles, notwithstanding the admirable and sometimes desirable effects associated with its invocation. As discussed in subsequent parts of this book, although it

remains a potential defense against causes of action to recover debts owed, the notion of odious debt is likely to be viewed as superseded by other more conventional legal arguments. The opportunity to closely examine this notion in the context of an actual real-life situation in which it has been raised is far too enticing to forego. After all, if that odious debt doctrine has sufficient credence to earn a position in the general corpus of international law, it remains available as a supplementary principle to bolster other legal arguments for dismissing causes of action to recover Iraqi debts.

1 A PRÉCIS OF IRAQ'S DEBTS

I. INTRODUCTION

When considering the matter of debts owed by one sovereign nation-state to other entities, the inclusion of more than only payments contractually obligated for remittance to other autonomous states must be acknowledged. Indubitably, debts that are owed include these contractual commitments entered into with other states; however, they also encompass potential remittances associated with obligations that arise from the failure to meet the legal duties of commercial undertakings or general civil responsibilities – whether the contractual commitments, commercial obligations, or civil responsibilities involve governmental actors or private parties. That is, a nation-state may have borrowed money from a foreign nation and be contractually obligated to make payments of principal and interest to the lender; or it may have entered into a commercial arrangement with a foreign nation or private entity to purchase goods or secure certain services; or it may have faced and breached a specific duty or behavior owed to such nation or entity and thus be indebted to make payment for its commercial undertaking or civil infraction. The government of Iraq has debts owed to others that involve each of these forms of obligation.

This book, however, does *not* include an examination of Iraqi debts that may have arisen from civil infractions involving individuals subjected to mistreatment by Iraqis operating under color of governmental authority. A prominent example is the consolidated cases recently decided by the U.S. Supreme Court styled as *Simon v. Republic of Iraq*, in which compensation was sought for harm said to have resulted from the detention, physical deprivation, and severe mistreatment accorded

individuals captured and imprisoned by Iraqi military and intelligence forces during the First Gulf War in 1991.[1] When these cases are instituted in U.S. courts, they necessarily implicate those aspects of the Foreign Sovereign Immunities Act (FSIA) that are designed generally to permit causes of action to proceed – despite the fact of filing against another nation-state – whenever that state is designated a sponsor of terrorism and has inflicted the personal injury leading to the lawsuit through actions amounting to torture or hostage taking.[2] These cases raise numerous complexities, and necessarily concern debts of a substantially uncertain nature, given both the undetermined number of individuals who may wish to pursue such claims and the unclear nature and level of compensation to which they may be deemed entitled.[3]

Regarding the Iraqi debts of relevance to this book – that is, those that are a result of borrowing or are of a contractual or commercial nature – at the time of the Second Gulf War and Saddam Hussein's removal from power, levels of total Iraqi debt reportedly ranged from $120 billion (USD) to nearly $400 billion. However, these estimates included a substantial sum in claims still pending before the United Nations Compensation Commission (UNCC)[4] in connection with Iraq's invasion of Kuwait in August 1990 and the First Gulf War. To illustrate the range of estimates, information from the World Bank in 2003 indicates that Iraq faced considerable reconstruction efforts

[1] See "Court Steps into Case Involving Iraq: Justices Intervene in Suit Against Iraq for Actions Under Saddam," Jan. 9, 2009, available at www.cbsnews.com/stories/2009/01/09/ap/supremecourt/main4710562.shtml (accessed Jan. 12, 2009). For the opinion delivered on the case by the U.S. Court of Appeals for the D.C. Circuit, see *Robert Simon et al. v. Republic of Iraq et al.* (decided June 24, 2008), available at http://caselaw.lp.findlaw.com/data2/circs/dc/067175p.pdf (accessed Jan. 12, 2009).

[2] See especially 28 U.S.C. Sec. 1605(a)(7).

[3] For other similar cases, see *Beaty v. Republic of Iraq*, 480 F. Supp. 2d 60 (D.C.Cir. 2007); *Vine v. Republic of Iraq*, 459 F. Supp. 2d 10 (D.C. Cir. 2006); *Acree v. Republic of Iraq*, 370 F. 3d 41 (D.C.Cir. 2004); and *Daliberti v. Republic of Iraq*, 97 F. Supp. 2d 38 (D.C.Cir. 2000).

[4] The report of Frederick D. Barton and Bathsheba N. Crocker, infra note 6, which fixed Iraqi debt at the time of Saddam's removal at nearly $400 billion, estimated that this total included more than $170 billion in UNCC claims not settled as of the time of the estimate itself. Subsequent thereto, those additional claims have been settled and awards made amounting to only $10 billion or so. This figure is derived from the fact that the Barton and Crocker study places decided UNCC claims at the time of Saddam's removal at $43 billion; see id. at note 1. The official UNCC Web site information indicates that at the conclusion of its work, awards totaled $52.4 billion; see UNCC, infra note 9.

and required a one-third forgiveness of the $120 billion debt load.[5] An earlier 2001 report from the World Bank determined the debt level to be more than $127 billion.[6] Furthermore, a 2003 estimate by the Center for Strategic and International Studies (CSIS) reported the total Iraqi debt – including pending loan repayment, commercial commitments at the time of Saddam Hussein's removal, preexisting contractual financial obligations, and First Gulf War compensation (including personal injury and property loss) – at $383 billion.[7] It is clear that in arriving at each estimate, the best available – yet admittedly imperfect – information was used by the sources. There have been indications that until a 1991 report to the United Nations (UN) Security Council by Iraqi representatives, official statistics on Iraq's total external debt load of more than $42 billion – excluding interest and funds denoted by Iraq as "grants" from other Gulf States – had never been released by the Iraqi government.[8]

At least two factors are important in determining Iraq's total debt load as of early 2009. First, all claims associated with the work of the UNCC have now been resolved (although disbursement of all payments may not have occurred yet),[9] resulting in a downward revision of the highest original estimate of nearly $400 billion to just under $200 billion.[10] Second, as discussed herein, Iraqi creditor nations that

[5] See United Nations/World Bank Joint Iraq Needs Assessment (Oct. 2003), available at http://siteresources.worldbank.org/IRFFI/Resources/Joint±Needs±Assessment.pdf (accessed Jan. 12, 2009); and Martin Crutsinger, "World Bank Calls on Rich Countries to Forgive One-Third of Iraq's Debt," Associated Press, Oct. 29, 2003.

[6] See Frederick D. Barton and Bathsheba N. Crocker, "A Wiser Peace: An Action Strategy for a Post-Conflict Iraq, Supp. I: Background Information on Iraq's Financial Obligations" at 4 (Jan. 23, 2003), available at www.csis.org/media/csis/pubs/wiserpeace_i.pdf (accessed Jan. 9, 2009).

[7] See id. at 1.

[8] See Ahmed M. Jiyad, "An Economy in a Debt Trap: Iraqi Debt 1980–2020," *Arab Stud. Q.* (Fall 2001), available at www.accessmylibrary.com/coms2/summary-0286-27240808_ITM (accessed Jan. 13, 2009).

[9] See UNCC, "At a Glance," available at www.uncc.ch/ataglance.htm (accessed Jan. 19, 2009).

[10] Indications are that the UNCC concluded its work and awarded relief on approximately 1.55 million claims, with monetary relief totaling $52.4 billion. See id. As of the end of 2008, more than $26 billion had already been paid by the UNCC, with another $26 billion or so yet to be paid. When the highest original estimate of nearly $400 billion was made, it was arrived at by including a $199 billion estimate for claims processed by the UNCC. That figure was based on well in excess of $300 billion in several categories of claims originally submitted to the UNCC, of which two categories totaling $148 billion

are members of the so-called Paris Club – that is, the major capital-exporting states – completed a process of forgiving 80 percent of the debt owed by Iraq. In addition to these two significant developments, the discussion that follows describes other efforts – at both private and governmental levels – that have further contributed to the reduction of Iraq's debt load. Although this book does not focus on more than a nonspecific and general picture of Iraq's debt obligation, the combination of these factors has reduced the debt to creditor states and private parties to a more manageable level.

For the purpose of discussion, we start from the assumption that when the Second Gulf War concluded, the CSIS debt estimate of $383 billion was the most accurate and all-inclusive and would be distributed among various types of claimants. Approximately $57 billion would be associated with pending commercial obligations for goods and services scheduled to be provided to Iraq. Another $127 billion would be in the form of existing contractual debt deriving from loans and other borrowing activities of the Iraqi government as well as commercial-credit purchases. Approximately $199 billion of Iraq's debt would be related to UNCC claims. As previously indicated, however, the UNCC claims process has concluded and the original $199 billion resulted in only $52.4 billion in awards, with $26 billion of that amount yet to be paid.[11] Similarly, all but 20 percent of the contractual and commercial debt otherwise owed by Iraq to Paris Club governments has been forgiven (the United States even forgave the additional 20%), thereby accounting for a further reduction of the original debt figure by approximately $45 billion.[12] It is obvious, however, that not all of Iraq's original contractual and commercial debt was owed to Paris Club states. Of those other states and private enterprises that are owed money, billions of dollars in claims are apparently still outstanding, although

had been settled for approximately $43 billion as of Saddam's removal. Of that $43 billion, only $16 billion had been paid, leaving about $27 billion outstanding, plus roughly $175 billion to $200 billion in two other claims categories that had not been addressed as of that time. See Frederick D. Barton and Bathsheba N. Crocker, supra note 6 at 1 (setting figure at $320 billion); and UNCC, "Status of Processing and Payment of Claims," available at www.uncc.ch/status.htm (accessed Jan. 21, 2009) (setting figure at $352+ billion).

11 See supra note 10.

12 See Karrim Yasser, "Paris Club Cancels $45 Billion of Iraq's Debt," Iraq Updates (6 Jan. 2009), available at www.iraqupdates.com/p_articles.php/article/42895/refid/MR-story-LM-13–01–2009 (accessed Jan. 20, 2009).

efforts to reduce private and non–Paris Club state debt have been successful.

II. UNCC DEBTS: RESOLVED AND HALF PAID

As previously discussed, UNCC claims have been resolved (although only about half paid). However, it is interesting that the officially reported $52.4 billion final total of awards began with the amount for all resolved claims at more than $350 billion.[13] Claims for payment considered by the UNCC were divided into six distinct categories and labeled Categories A through F; the first four categories correspond roughly to claims by individuals and families and the other two to claims by companies, governments, and international organizations.[14] The UNCC permitted only governments and international organizations to submit claims, thereby requiring individuals, families, and companies initially to process claims through a fact examining and screening mechanism established by their own government.[15]

In terms of claims resolved, more than 2.6 million of the nearly 2.7 million total were in Categories A, B, C, and D (i.e., individuals and families).[16] Approximately seven thousand claims were from Categories E and F claimants (i.e., companies, governments, and international organizations).[17] Although claimants in the first four categories accounted for the majority of all claims resolved, they amounted to only about 20 percent of the total $52.4 UNCC awards. Specifically, although there were approximately 2,800 claims from companies in Category E, the awards granted amounted to more than $26 billion.[18] Likewise, the awards for the approximately 4,100 claims from governments and international organizations totaled more than $14 billion.[19] As a result of the Kuwait invasion and consequent First Gulf War, it is

[13] See UNCC, "Status of Processing and Payment of Claims," available at www.uncc.ch/status.htm (accessed Jan. 21, 2009).

[14] See UNCC, "The Claims," available at www.uncc.ch/theclaims.htm (accessed Jan. 21, 2009).

[15] See generally Nicholas C. Ulmer, "Claimant's Expectations from the United Nations Compensation Commission," 15 *J. Int'l Arb.* 7 (Mar. 1998).

[16] See UNCC, "Status of Processing and Payment of Claims," supra note 13.

[17] See id.

[18] See id.

[19] See id.

more likely that companies, governments, and international organizations were substantially more at risk than individuals; therefore, it may be better to view the awards granted by the UNCC relative to the percentage of returns on the amount of compensation originally claimed. Using that measure, individuals originally claimed about $31.5 billion and received awards of $11.8 billion,[20] for a return of better than 33 percent. Conversely, companies, governments, and international organizations received awards exceeding $40 billion, a more than 12 percent return on their original claims of approximately $320 billion.[21]

Reviewing the nations represented by the claimants in the six categories recognized by the UNCC, more than 300,000 of the roughly 923,000 resolved Category A claims were Egyptian nationals and more than 112,000 claims were from Indian nationals. Collectively, these two countries accounted for about 40 percent of all Category A claims. Claims also were considered from individuals and families from nations in the vicinity of Kuwait, including Iran, Jordan, and Syria, as well as from Indo-Pacific countries including Sri Lanka, Bangladesh, and the Philippines.[22] Approximately 5,700 resolved claims were from Category B claimants, with many of the same Category A nationalities represented. However, more than 50 percent involved Kuwaitis, with a total of approximately three thousand claims. There were also fifty-nine Category B claims from U.S. claimants.[23] Category C represented the single largest group with approximately 1.7 million claims. Officially, Egypt submitted more than 1.2 million claims; when added to the more than 162,000 claims of Kuwaiti origin, they comprise 75 percent of the claims in Category C.[24] Category D involved more than 5,500 claims by Kuwaiti nationals, again approaching 40 to 45 percent of the more than 13,000 claims in that category, which also had 317 claims from U.S. individuals and families.

In Categories E and F (i.e., company, government, and international organization claims), approximately 6,700 Category E claims

[20] See id.

[21] See id.

[22] See UNCC, "Category 'A' Claims," available at www.uncc.ch/claims/a_claims.htm (accessed Jan. 21, 2009).

[23] See UNCC, "Category 'B' Claims," available at www.uncc.ch/claims/b_claims.htm (accessed Jan. 21, 2009).

[24] See UNCC, "Category 'C' Claims," available at www.uncc.ch/claims/c_claims.htm (accessed Jan. 21, 2009).

submitted to the UNCC were resolved. Of those, roughly 2,800 (or slightly more than 40% of the total) involved injury or loss suffered by the government of Kuwait, with another 152 claims involving the U.S. government. Other claims were submitted by Israel, France, Egypt, India, Italy, the Netherlands, Germany, Turkey, the United Kingdom, and Saudi Arabia.[25] The UNCC processed more than 380 Category F claims, of which 70 were from Kuwait and 60 were from Saudi Arabia. In total, forty-three governments and six international organizations submitted Category E claims.[26]

As discussed previously,[27] only about $26 billion of the UNCC-awarded $52.4 billion had been paid to successful claimants as of early 2009. The principal reason for this is that monies to fund the awards come from an earlier UN-mandated requirement that Iraq ensure that 5 percent of oil and gas revenues be available for Kuwaiti war claims. The result is that approximately half of the UNCC-resolved awards remains an outstanding debt obligation of the Iraqi government. The Iraqi government has indicated that it wants to jettison the requirement to continue making contributions to fund the compensation claims from the First Gulf War;[28] however, even setting aside that debt, it still faces the potential for considerable outstanding financial claims from various sources. As long as the sale of Iraqi crude oil continues to be the main source of revenue for Baghdad, there is a distinct risk that the resources and the monies they generate could become the object of legal actions initiated by creditor claimants.

III. CONTRACTUAL DEBT FROM BORROWING AND COMMERCIAL PURCHASING ACTIVITY

Thinking in terms of the CSIS estimate that total debt owed by Iraq amounted to $383 billion at the time of the Second Gulf War and the removal of Saddam Hussein from power, the resolution of all UNCC

[25] See UNCC, "Category 'D' Claims," available at www.uncc.ch/claims/d_claims.htm (accessed Jan. 21, 2009).

[26] See UNCC, "Category 'E' Claims," available at www.uncc.ch/claims/e_claims.htm (accessed Jan. 21, 2009).

[27] See supra note 10.

[28] See "Iraq Calls for End to UNCC," Jubilee Iraq (June 28, 2005), available at www.jubileeiraq.org/blog/2005_06.html#000845 (accessed Jan. 21, 2009).

claims suggests reducing that estimate nearly to half (because it was premised on then-outstanding claims of $172 billion – since resolved – and unpaid awards totaling $27 billion).[29] Of the remaining $180 billion to $200 billion – that is, claims not within the UNCC compensation process – a considerable portion derives from borrowing and commercial purchasing activity of the various Iraqi government entities and affiliated financial and business groups.

By the CSIS estimate, at the time of the Second Gulf War, Iraq had a debt burden from its borrowing and credit-purchasing activities of approximately $127 billion.[30] The CSIS estimate acknowledged, however, that the estimates reported by others ranged from $62 billion to $130 billion.[31] Admittedly, there always has been some controversy between Iraq and many of the neighboring Gulf states over the precise nature and character of monies extended by those states to the Iraqi government. In the Iraqi government's view, a substantial portion of the borrowed money was offered in the form of "grants" that need never be repaid rather than loans with a fixed repayment obligation.[32] Even beyond that fundamental point, it would be seriously remiss not to call attention to the fact that as much as $47 billion of the $127 billion owed by Iraq at the time of the Second Gulf War resulted from accrued interest on outstanding principal.[33] The passage of several more years has surely considerably increased the amount of interest owed.

What about the form of the borrowing or credit-purchase obligations that resulted in the original Iraqi debt? According to knowledgeable sources that closely examined the issue, it appears to be primarily in the form of short-term debt owed to private banks and companies.[34] That is not to say that longer-term debt is not owed to foreign governmental institutions and regional and international lending agencies.

[29] See Frederick D. Barton and Bathsheba N. Crocker, supra note 6 at 1.
[30] See id.
[31] See id. at note 3.
[32] Using different numbers, but raising exactly the point suggested, see Ahmed M. Jiyad, supra note 8 at 6.
[33] See Frederick D. Barton and Bathsheba N. Crocker, supra note 6 at 3, figure 3.
[34] See id. See also Ahmed M. Jiyad, supra note 8 at 3–5, and "The Cost of War and Reconstruction, Iraqi Debt – Who Will Foot the Bill?" 46 *Middle East Economic Survey* (No. 10) (Mar. 10, 2003), available at www.mees.com/postedarticles/finance/iraq46n10b01.htm (accessed Jan. 21, 2009).

Rather, most of Iraq's borrowing and purchasing debt due or pending at the time of the Second Gulf War was of a form that would mature rapidly and carry interest payments at higher rates than those associated with its borrowing and purchasing debt owed to foreign governments and regional and international banks.

A breakdown of Iraqi borrowing and purchasing debt by geographically situating relevant private or governmental creditors extant in the weeks and months prior to Saddam Hussein's removal is interesting and revealing. According to the CSIS, the largest debt was owed to the various Gulf States, with the 2002 accumulated amount at more than $30 billion. Iraq's debt to Kuwait is believed to have accumulated as early as 1992 (obviously from pre–First Gulf War obligations coming due) and surpassed $17 billion. With approximately $12 billion accumulated as of 2002, Russia ranks third among Iraq's creditors. Iraq has a debt to Bulgaria of $1 billion, accumulated since 1998, as well as a host of other creditor countries ranging in rank order from Turkey to Poland, Jordan, Morocco, Hungary, Egypt, and others. Again, it must not be overlooked that as of 2003, some sources argued that accrued interest alone on Iraq's outstanding debt was $47 billion.[35]

For a sense of the private banks and companies that Iraq reportedly owes money to as a consequence of borrowing and purchasing obligations, recall the *Commercial Bank of Kuwait* case referenced in the preface of this book.[36] Beyond that case, however, there are many others, some that even filed suit against Iraq or one of its agencies as in the *Commercial Bank* case. For example, *First City, Texas Houston Bank v. Rafidain Bank*, a court action commenced in November 1990 that involves a bank located in the United States, sought to recover $49.9 million in drafts acquired under letters of credit issued by Rafidain Bank and backed up by the CBI. Although First City was awarded a default judgment of $53.2 million against Rafidain, the case against the CBI and for collection of the default judgment has had various iterations in subsequent years.[37]

[35] See Frederick D. Barton and Bathsheba N. Crocker, supra note 6 at 3, figure 3.

[36] See preface at note 15.

[37] See e.g., *First City-Texas Houston, N.A. v. Central Bank of Iraq*, 150 F.3d 172 (2d Cir. 1998); *First City-Texas Houston, N.A. v. Rafidain Bank, Central Bank of Iraq*, Docket No. 00–9556 (2d Cir. 2002), available at http://caselaw.lp.findlaw.com/cgi-bin/getcase.pl?court=2nd&navby=case&no=009556 (accessed Jan. 22, 2009).

Focusing only on the private banks and financial companies that reportedly loaned money to Iraq in the past, the most prominent include the U.S.–headquartered Chase Manhattan (now J.P. Morgan Chase) and Irving Trust (now Bank of New York). Either directly or as a result of debt-trading activity, other financial institutions also are said to hold debt generated by Iraqi borrowing, including Emergent Alternative Fund, Aberdeen Asset Management, and Gramercy Advisors. The better-known overseas banks said to be owed money by Iraq include BNP Paribas, Banco Nazionale de Lavoro (BNL; now a subsidiary of BNP Paribas), British Arab Commercial Bank (BACB), Arab International Bank–Bahrain (ALUBAF), Commerzbank A.G., The Arab Investment Company–Bahrain (TAIC), and Union de Banques Arabes et Francaises (UBAF).[38] Debt originally owed to any of those creditors may well have been sold or traded to others because it is evident from the passage of time that Iraqi authorities were uninterested or incapable of meeting that nation's payment obligations. Consequently, actual claimants, either after the Second Gulf War or currently, may not parallel those involved in the initial lending activity precipitating the debt. Estimates at the time of Saddam Hussein's removal from power placed Iraq's borrowing from private banks and financial companies at $2.6 billion.[39] Whereas indications are that loans of substantially greater size were made to Iraq as early as 1983, approximately $32 million of the $2.6 billion was reportedly held by U.S. banks, according to a Federal Financial Institutional Examination Council study.[40] Those figures obviously may omit interest accumulated on the original loans.

Iraq also incurred borrowing debt with governmental and regional or international lending agencies. Reports indicate that perhaps the

[38] See "Saddam's Debts" (latest estimates), Jubilee Iraq (as of 2003) available at www.jubileeiraq.org/debt_today.htm (accessed Jan. 26, 2009). See also Chana R. Schloenberger, "Gambling on Ira's Reconstruction," Forbes.com (Apr. 4, 2003) ($500 million loan in 1983 by a syndicate of J.P. Morgan, Bank of New York, Irving Trust, and BNP Paribas), available at www.forbes.com/2003/04/04/c2_cs_0404iraqdebt.html (accessed Jan. 25, 2009); Cara Marcano, "Betting on Debt: Will Iraq Pay Up?" *Int'l Herald Tribune* (Apr. 26, 2003), available at www.iht.com/articles/2003/04/26/mfunds26_ed3_.php (accessed Jan. 26, 2009).

[39] See "Breakdown of Iraq's Foreign Debt," infra note 41; "Iraq's $326 Billion Debt, War Claims May Complicate Rebuilding," Bloomberg.com (May 8, 2003), available at www.bloomberg.com/apps/news?pid=nifea&&sid=ajN4f0LBBo8 (accessed Jan. 30, 2009).

[40] See "Iraq's $326 Billion Debt, War Claims May Complicate Rebuilding," id.

largest single debt was owed to the Gulf Cooperation Council states: Saudi Arabia, Kuwait, Bahrain, Oman, Qatar, and the United Arab Emirates (UAE). As alluded to previously, Iraqi authorities have argued not only that the roughly $30 billion received was in the form of grants rather than loans but also that estimates as high as $55 billion or more reflect exorbitant interest charges on any confirmed loans.[41] Also reported was that the Arab Fund for Social and Economic Development extended $166 million in assistance to Iraq in 1998, the Islamic Development Bank loaned $125 million in 1999, and the Arab Monetary Fund loaned $420 million in 2004.[42]

When added to the then-outstanding obligations to the Gulf Cooperation Council states, the Arab Fund for Social and Economic Development, the Islamic Development Bank, and the Arab Monetary Fund, the $2.6 billion owed by Iraq to private banks and financial companies as of the Second Gulf War collectively totals approximately $33 billion to $57 billion. Using the CSIS estimate of $127 billion for Iraq's total borrowing and commercial purchasing debt as of the Second Gulf War suggests that between $70 billion and $90 billion was in the form of accruals of interest and debt-servicing charges on loans from private banks and other governmental financial institutions, as well as unmet payments on commercial-purchasing obligations.

IV. COMMERCIAL DEBT FROM PENDING GOODS AND SERVICES CONTRACTS

In addition to not meeting payment obligations on loans received from private banks and financial lenders; to governmental, regional, and international lending agencies; and failing to meet payment schedules associated with goods and services acquired from suppliers, Iraq faced potential claims of debt related to contracts for goods and services that were formalized and pending at the time of the Second Gulf War, but on which Iraq reneged due to a change of heart or a distressed economic position. Claims of this nature with both private and governmental suppliers were reported by CSIS to reach the level of $57 billion.[43] This

[41] See "Breakdown of Iraq's Foreign Debt," infra note 43.
[42] See "Saddam's Debts," supra note 38.
[43] See Frederick D. Barton & Bathsheba N. Crocker, supra note 6 at 1.

represented roughly one-fourth of the estimated $383 billion total of Iraqi debt – the balance from $127 billion in borrowing and commercial purchases on credit terms, with $199 billion in then-unresolved UNCC claims related to the First Gulf War.

The $57 billion for pending contractual commitments was 90 percent controlled by Russian entities believed to hold as much of $52 billion in executed but pending obligations to supply goods and services.[44] In terms of other countries with pending contracts, the next single largest placeholder was the Netherlands, with $3.6 billion, or 6 percent of the total. Egypt, the UAE, China, and France followed with $740 million, $703 million, $80 million, and $73 million, respectively, in pending contracts at the time of the Second Gulf War.[45] It is clear that one reason for such significant amounts in pending contracts held by Russia, France, and the other placeholders was related to their historical linkage and affiliation with Baghdad. Another important reason, however, that was surely operative in the case of Russia and France was their increasing displeasure with the First Gulf War UN sanctions regime and their interest in getting ahead of others on Iraqi oil and gas opportunities.

The potential claims associated with pending contracts with Iraq could run the full range of goods and services traded on the international market and consumed in perhaps the oldest civilization on earth. Although that may be a generally accurate assumption, at least one evaluation indicated that $38 billion of the $57 billion total in pending contracts was related to oil-sector goods and services.[46] The CSIS study indicated that telecommunications was another major sector involving pending contracts.[47] Indeed, it was reported that Chinese entities were poised to provide Iraq with the benefits of a substantial fiber-optics cable project at the time of the Second Gulf War.[48] Additionally, the French motor-vehicle companies Peugeot and Renault and the telecommunication provider Alcatel were believed to have lucrative pending contractual commitments to deliver minibuses, garbage trucks, and telecom equipment.[49] It is highly unlikely that these

44 See id. at 4.
45 See id.
46 See Colum Lynch, "Iraq Uses Its Buying Power as Leverage," *Wash. Post*, July 3, 2001.
47 See Frederick D. Barton & Bathsheba N. Crocker, supra note 6 at 4.
48 See Colum Lynch, supra note 46.
49 See id.

and many other public or private enterprises around the globe were prepared to accept the complications associated with the radical governmental changes resulting from Saddam Hussein's ouster and to turn their back on signed and sealed ironclad business commitments solemnly undertaken at the national level. It is understandably difficult for replacement or successor governments to scrape together the capital needed to meet their nation's contractual commitments. However, in the absence of an Iraqi willingness to satisfy its side of the bargain, the option by the beneficiaries of such business arrangements for recourse to appropriate legal processes was an important cudgel to retain.

V. CONCLUSION

Other than debt related to the UNCC claims process, debts associated with Iraqi borrowing and completed credit-purchase activity, as well as pending contracts for goods and services from various providers, were considerably reduced through negotiated efforts, with speculation that the total debt load *remaining as of the beginning of 2009* was between $50 billion and $70 billion.[50] A study completed by the International Monetary Fund (IMF) in 2007 hoped that by 2009 the total would be closer to $34 billion to more than $62 billion.[51] Efforts to secure reduction to this more manageable level were carried out through several groups: the Paris Club, the London Club, the Iraqi Creditors Club, and the Iraq Debt Reconciliation Office (IDRO), headquartered in Amman, Jordan. One analysis of the remaining 2009 debt load suggests that the Paris Club states were owed $7.6 billion; the Gulf Cooperation Council nations of Saudi Arabia, Kuwait, Bahrain, Oman, Qatar, and the UAE were owed between $30 billion and $50 billion;

[50] See "Breakdown of Iraq's Foreign Debt," PUKmedia (26 May 2008), available at http://pukmedia.com/english/index.php?option=com_content&task=view&id=4491&Itemid=1 (accessed Jan. 26, 2009) (fixing total between $56.6 billion and $79.9 billion, but subsequently, the UAE canceled the $7 billion owed to it). On the UAE cancellation, see "UAE to Cancel Iraq's $7 Billion Debt" (July 6, 2008), available at www.cnn.com/2008/WORLD/measr/07/06/iraq.uae/index.html (accessed Jan. 28, 2009).

[51] See IMF, "Iraq: 2007 Article IV Consultation, Fifth Review Under the Stand-By Arrangement," *IMF Country Report* No. 07/301 (Aug. 2007), at 37, Tbl.1, "Iraq: Estimated External Debt Stock, 2004–12."

non–Paris Club states and governmental or intergovernmental entities with bilateral agreements with Iraq were owed approximately $640 million; and other entities are owed in the vicinity of $16 billion.[52] Although difficult to develop a definitive picture, new reports leading up to 2010 suggest the Chinese, in return for favorable consideration of their interest in Iraqi oil fields, consented to reduction on the more than $8 billion debt owed them, whereas debt owed to Kuwait remained under reduction pressure from the United States and others.[53]

From the standpoint of news-media coverage of Iraqi debt reduction, the Paris Club efforts clearly received the most extensive attention. As described previously, the Paris Club is an informal group of nineteen creditor governments, primarily from industrialized and capital exporting nations: Australia, Austria, Belgium, Canada, Denmark, Finland, France, Germany, Ireland, Italy, Japan, the Netherlands, Norway, Russia, Spain, Sweden, Switzerland, the United Kingdom, and the United States. First established in 1956, it attempts to reschedule or cancel financial obligations of highly indebted poor nations. In the case of Iraq, the process began in summer 2004 with the so-called Evian G-8 Summit, which addressed many topics including the matter of debt forgiveness for a resource-rich country like Iraq.[54] Led by the United States, the effort culminated with a November 2004 Paris Club agreement.[55] Although several Paris Club member states were reluctant to be too aggressive in forgiving Iraqi debt, a reduction of total debt owed to member governments by a factor of 80 percent was eventually settled. This was delivered in three tranches: the first two in 30-percent increments dependent on Iraq satisfying specific

[52] See "Breakdown of Iraq's Foreign Debt," id.

[53] On the matter of China and debt, see "China Slashes Iraq Debt for Oil Deals," Upstreamonline.com (Nov. 10, 2009), available at www.upstreamonline.com/live/article198380.ece (accessed Jan. 20, 2010); "Iraq Says China Agrees to Forgive 80 Pct of Debt," (Nov. 10, 2009), available at www.iraqdirectory.com/Display/News.aspx?id=10832 (accessed Jan. 20, 2010). With respect to Kuwait, see "Iraq Seeks Forgiveness of Saddam Hussein's 'Odious' Debts," ProbeInternational (June 21, 2009), available at www.probeinternational.org/odious-debts/iraq-seeks-forgiveness-of-saddam-hussein's-odious-debts (accessed Jan. 20, 2010); "Iraq and the Gulf on the Rocks," Foreign Policy (June 2, 2009), available at http://lynch.foreignpolicy.com/posts/20099/06/02/iraq_and_the_gulf_on_the_rocks (accessed Jan. 20, 2010).

[54] See Martin A. Weiss, "Iraq's Debt Relief: Procedural and Potential Implications for International Debt Relief," CRS Report to Cong., Order Code RL33376 (Oct. 2, 2008) [hereinafter, CRS Report] at 6.

[55] See id. at 7.

benchmarks and the final 20 percent awarded during summer 2008.[56] As mentioned previously, the United States acted more aggressively and forgave Iraq the remaining balance of its debt,[57] although speculation in late December 2008 was that Paris Club states were still owed, as previously indicated, more than $7 billion.[58] Given that Iraqi debt was previously separated into UNCC, borrowing, and credit purchases, as well as pending-contract debts, some of the more than $7 billion owed to Paris Club states was probably distributed between the latter two categories. Presumably, Paris Club states exist who still expect payment from Iraq on its remaining unforgiven obligation.

Iraqi debt owed to private banks and lending entities was addressed initially by the so-called London Club; however, as discussed herein, it seems to have generally collapsed into negotiations with all other private commercial creditors, most notably those holding debt for sales or services provided to Iraq, including pending contracts. The London Club, which dates from the mid-1970s, represents private commercial banks headquartered in capital-exporting nations. Early indications suggested debt reduction negotiated through the efforts of the London Club might result in Iraq receiving a better deal than through the Paris Club, for two reasons: (1) Paris Club states had bound Iraq to a commitment to not negotiate with non–Paris Club entities on terms more favorable than given to Paris Club creditors; and (2) by virtue of their private nature, the commercial banks and lending institutions did not possess the same leverage that the official governmental institutions had that negotiated the Paris Club agreement.[59]

By the end of April 2004, the British financial and debt-negotiation advisors GML, Ltd., and Exotix collaborated on launching an effort to organize creditors that were owed money by Iraq as a result of commercial trade and services activities.[60] The organization, named the Iraq Creditors Club, had been preceded in late 2003 by Hyundai's

56 See id.

57 See id. at 7–8.

58 See "Paris Club Lenders Complete Iraq Debt Forgiveness," Project Rebuild Iraq 2008 (Dec. 29, 2008), available at www.rebuild-iraq-expo.com/newsdetails05,asp?id=4554 (accessed Jan. 26, 2009).

59 See, generally, Joanna Chung, "Iraq's Debt Solution Ruffles Feathers," *Financial Times* (Dec. 21, 2005), available at www.probeinternational.org/odious-debt/news-and-opinion/iraqs-debt-solution-ruffles-feathers (accessed Jan. 30, 2009).

60 See "Iraq Creditors Club Launched" (Apr. 30, 2004), available at www.emta.org/ndevelop/Iraq_Creditors_Club_Launched.pdf (accessed Jan. 30, 2009).

efforts to create the so-called Washington Club, which was composed of about twenty creditors from eight nations that aimed to collectively press their claims for payment on past debt.[61] To address the matter of all private commercial claims (whether deriving from Iraqi borrowing, trade, or contracting activity – without distinction between performed and pending contracts), the IDRO was established in Amman, Jordan, in May 2004 by the new government in Baghdad.[62] Formally, IDRO operates under the Iraqi Ministry of Finance; actually, much of its work is performed by the accounting and financial experts of Ernst & Young. In December 2004, the Ministry of Finance indicated that it would publicly call for all commercial claimants to present claims to IDRO for resolution; on July 26, 2005, it announced the settlement terms.[63] Those terms envisioned only nongovernmental commercial claimants, including those holding debt deriving from financial or contractual obligations extant as of the First Gulf War (i.e., as of August 6, 1990).[64] Furthermore, to the extent that any relevant claim did not exceed $35 million, it was to be settled at an exchange of 10.25 percent on the dollar and resolved through a cash payment. Claims in excess of $35 million were to be settled at no more than 20 percent on the dollar and resolved through either a syndicated loan or a privately issued bond scheduled to mature in 2028 and bearing a 5.8 percent interest rate.[65]

On September 16, 2005, IDRO announced the conclusion of the opening phase of its reconciliation effort and that of the roughly $750 million in small claims submitted and deemed eligible, 78 percent of the claimants accepted the proposed settlement amounts.[66] Fewer than three months later, in December 2005, IDRO announced a proposed debt-for-debt exchange envisioned for $14 billion of larger claims, which represented about 60 percent of the total demanded

[61] Available at www.jubileeiraq.org/blog/ (Click on "Private" in the right-hand column and scroll down to Oct. 10, 2003, report).

[62] See CRS Report, supra note 43 at 11.

[63] See "Iraq Announces Terms of Commercial Debt Settlement Offer" (July 26, 2005), available at www.eyidro.com/iraqterms.pdf (accessed Feb. 4, 2009).

[64] See id.

[65] See id.

[66] See "Iraq Announces the Successful Conclusion of First Phase of Commercial Debt Restructuring" (Sept. 16, 2005), available at www.eyidro.com/pr-conc.pdf (accessed Feb. 4, 2009).

by all eligible claimants that had registered for debt reconciliation.[67] On January 11, 2006, Iraq confirmed the acceptability of settlement offers regarding those claims, noting that the $14 billion was rescheduled at an amount no greater than $2.7 billion.[68] At the end of that month, IDRO issued a notice indicating that it had closed with the larger creditors on debt-for-debt exchanges, rescheduling the $14 billion that had been claimed.[69] By spring 2006, the results of IDRO's entire commercial-debt reconciliation exercise were reported: $18.4 billion in initial eligible claims were resolved, thereby canceling 10,247 claims of 456 participating claimants that held 96 percent of the value of eligible and submitted claims. An additional 825 claims were submitted to arbitration before three distinguished and internationally recognized arbitrators: E. Michael Hunter, Fathi Kemicha, and Ghaleb Mahmassani.[70] Most of the claims in arbitration were resolved quickly, with an announcement on June 20, 2006, that an additional $143 million in debt-for-debt exchange had been issued.[71] In mid-July of that year, IDRO issued another statement revising the results of its efforts: $19.6 billion of initial eligible claims had been resolved, which accounted for 11,776 individual claims submitted by 491 claimants, and 817 of the 825 arbitral proceedings also had been reconciled.[72] Although this announcement marked the successful conclusion of a commercial-claims process begun months before, it was recognized that many other

67 See "Iraq Announces Successful Conclusion of Debt-for-Debt Exchange Offer" (Dec. 23, 2005), available at www.eyidro.com/Iraq_Announces_Successful_Conclusion_of_Debt-for-Debt_Exchange_Offer.pdf (accessed Feb. 4, 2009).

68 See "Iraq Confirms the Results of Successful Debt-for-Debt Exchange Offer and Approval of IMF Standby Arrangement" (Jan. 11, 2006), available at www.eyidro.com/Iraq_Confirms_the_Results_of_Successful_Debt-for-Debt_Exchange_Offer_and_Approval_of_IMF_Standby_Arrangement.pdf (accessed Feb. 4, 2009).

69 See "Iraq Confirms the Closing of Debt-for-Debt Exchange Offer" (Jan. 23, 2006), available at www.eyidro.com/doc/Iraq_Confirms_Closing_of_Debt-for-Debt_Exchange_Offer.pdf (accessed Feb. 4, 2009).

70 See "Iraq Announces Results of Commercial Debt Settlement" (Apr. 3, 2006), available at www.eyidro.com/Iraq_Announces_Results_of_Commercial_Debt_Settlement.pdf (accessed Feb. 4, 2009).

71 See "Iraq Will Issue Additional Notes Under its Debt-for-Debt Exchange Offer" (June 20, 2006), available at www.eyidro.com/doc/Iraq_Will_Issue_Additional_Bonds.pdf (accessed Feb. 4, 2009).

72 See "Iraq Announces Conclusion of Commercial Debt Settlement" (Jul. 18, 2006), available at www.eyidro.com/doc/Iraq_Announces_Conclusion_of_Commercial_Debt_Settlement.pdf (accessed Feb. 4, 2009).

commercial claimants may not have participated. Thus, on January 30, 2008, IDRO announced the opening of another round in the claims process, with claims to be submitted for consideration by March 15, 2008.[73] In mid-May of that year, invitations were sent to those deemed eligible for settlement[74]; even as late as the beginning of 2010, results of that process had not been formally announced on IDRO's Web site.

There has never been any readily available public announcement identifying those who submitted claims to IDRO. Nevertheless, at least the nature of the enterprises engaged in by the petitioners would be suggested by the list of those known to have submitted claims to the U.K. court-established body set up to settle claims against the Iraqi government–controlled Rafidain Bank located in Britain and to resolve a legal dispute concluded prior to UN measures that insulated Iraq from court actions.[75] Of a list of approximately 140 claimants – all of whom were also informed by the Rafidain Bank administrators to submit claims to IDRO following its establishment – there were about twenty-five individuals and a number of estates of individuals; various banks and financial institutions including Bank of Tokyo, Citibank, DZ Bank Limited, Kuwait Financial House, and National Westminister Bank; and companies and enterprises including Apollo Fashions Co., Britannia Foods, Continental Construction Limited of New Delhi, Doosan Heavy Industries, HK Engineering, and Modern Construction Material Industry.[76] These entities certainly suggest that IDRO was petitioned for debt reconciliation by private citizens, major financial institutions, suppliers of basic consumer staples, and providers of heavy and sophisticated industrial products, as well as those with technical expertise and other services useful in various sectors of the economy.

Approximately $30 billion to $50 billion reportedly was owed as a consequence of monies earlier loaned to Baghdad by the Gulf

[73] See "Iraq Announces Limited Re-Opening of Commercial Debt Settlement Offer" (Jan. 30, 2008), available at www.eyidro.com/doc/1843505-1(Final688-08Re-openingPressRelease(30January2008)).PDF (accessed Feb. 4, 2009).

[74] See "Invitations Sent to Commercial Claimants" (May 16, 2008), available at www.eyidro.com/doc/1893853-3(688-08CommercialClaimsUpdate_1).PDF (accessed Feb. 4, 2009).

[75] See "Rafidain Bank (in Scheme of Arrangement)," available at www.rafidainbank.co.uk/News.aspx (accessed Feb. 5, 2009).

[76] See "Claims Submitted in the Scheme of Arrangement" (16 Oct. 2008), at id.

Cooperation Council states of Saudi Arabia, Kuwait, Bahrain, Oman, Qatar, and the UAE – monies that Iraq argued were "grants" rather than loans. Prime Minister al-Maliki was prepared to insist that it and other debts be canceled during a UN conference on Iraq scheduled for early summer 2008.[77] From news reports, after negotiations related to that conference, the UAE agreed to extinguish its claim for $7 billion.[78] Other Gulf Cooperation Council states indicated a move in the same direction. Saudi Arabia, Oman, and even Kuwait expressed an interest,[79] although they have been reluctant to convert intention into reality. It is obvious that in the context of sectarian political tensions that affect Middle East international relations, the Sunni-controlled regime in Saudi Arabia does not see eye to eye with Shiite-dominated Iraq. In the context of Kuwait, that country has the additional burden of having suffered the ravages and depredations of a brutal Iraqi military invasion, further complicating its willingness to follow through on expressions of flexibility regarding debt reconciliation or accede to overtures tending in that direction. In the end, it well may be that the substantial Gulf Cooperation Council debt is canceled or rescheduled to more manageable levels, consistent with the arrangement adopted by the Paris Club. Were that to happen, Iraq's non–UNCC debt load could be reduced to a few tens of billions of dollars, considerably more tolerable than what currently exists but still a potential source of major legal risks going forward.

[77] See "Iraq Seeks Foreign Debt Cancellation," CBS News (May 29, 2008), available at www.cbsnews.com/stories/2008/05/29/iraq/main4134289.shtml (accessed Jan 8, 2009).

[78] See "UAE to Cancel Iraq's $7 Billion Debt," supra note 50. See also "UAE Waives Billions of Iraqi Debt," BBC News (6 July 2008), available at http://news.bbc.co.uk/1/hi/world/middle_east/7492115.stm (accessed Jan. 9, 2009).

[79] See, e.g., "Kuwait to Ask MPs to Forgive Iraq Debt," ArabianBusiness.com (28 Aug. 2008), available at www.arabianbusiness.com/529297-kuwait-to-ask-mps-to-forgive-iraq-debt?In=en (accessed Feb. 1, 2009); "Arab Nations Fail to Forgive Iraq's Debts," *Associated Press* (May 30, 2008), available at http://query.nytimes.com/gst/fullpage.html?res=9401EED9103DF933A05756C0A96E9C8B63 (accessed Feb. 4, 2009).

2 THE LEGAL THEORY OF "ODIOUS DEBT"

I. INTRODUCTION

It is clear from the discussion in the preface that there is support in the basic notions of international law (e.g., *ex injuria*, *postliminium*, and the laws of war), and in that portion of national law regarding international transactions that become mixed up with military hostilities and occupation, generally preferring the claims of creditors to those of debtor states. As might be imagined regarding Iraqi debt, various theories and defenses can be advanced to blunt the attempts to work those notions so as to recover from Iraqi oil and gas – and revenues from their sale – what is owed to legal claimants. Almost immediately after the invasion of Iraq by coalition forces and the removal of Saddam Hussein from power, a particular defensive theory designed to insulate Iraq from creditors was voiced repeatedly: "odious debt." Given that the theory extends only to debts incurred by a previous regime for unsavory purposes, relative to Iraq, it was limited in its applicability to Saddam-era debts. In raising "odious debt," the idea was to shift the discussion dynamics and situate the claims for relief against Iraq in a legal milieu outside that which normally would expect creditors to have their claims fully satisfied.

The motivations associated with voicing the theory may be debated, but a number of different factors apparently contributed to that fact. First, as is well known, during the interwar years between the First and Second Gulf Wars, Iraq essentially had been under a UN-imposed trade embargo that seriously constrained its ability to generate revenues through oil and gas sales. With Saddam Hussein's removal from power, the embargo was lifted and the potential for a revenue stream

that creditors might view with interest was thus reinvigorated. Second, because of the horrible physical state of basic infrastructure and social services in the country after more than a decade of UN sanctions and depredations by Saddam's Baathist regime, it was essential that whatever revenues were generated by Iraq through sales of hydrocarbons be earmarked for national rebuilding, not the payment of creditor claims. Third, there was a sense that the U.S. efforts in the Second Gulf War had not been supported by certain European nations – particularly France, Germany, and Russia – to the extent that they (or private creditors resident therein) were owed monies by Iraq. Perhaps the invocation of any legal theory that might complicate recovery efforts would be effective in communicating the degree of U.S. dissatisfaction about having to largely "go it alone."

There were earlier intimations that the theory of odious debt should be applied to relieve Iraq of the burden of Saddam-era debts.[1] However, it was Paul Wolfowitz, Deputy Secretary at the U.S. Department of Defense and a principal architect of the war against Iraq, who at the beginning of the invasion suggested before the Senate Armed Services Committee the reliance on the essence of the theory. With specific reference to the debts owed Russia and France and their role in Iraq's future, he stated: "I hope, for example, they'll think about the very large debts that come from money that was lent to the dictator to buy weapons and to build palaces and to build instruments of repression. I think they ought to consider whether it might not be appropriate to forgive some or all of that debt so that the new Iraqi government isn't burdened with it."[2] Simultaneously, U.S. Treasury Secretary John Snow indicated to the media that "[t]he people of Iraq shouldn't be saddled with those debts incurred through the regime of the *dictator* who's now gone."[3]

The theory also was widely invoked beyond government circles. An editorial in the *Wall Street Journal* at the end of April 2003 suggested

[1] See, e.g., L. Solomon, "Iraq's Odious Debts," *National Post* (12 Feb. 2003).

[2] Quoted in Justin Alexander & Wajeeh Elali, "Iraq and the Odious Debt Doctrine" at 23 (2008) (draft of Cambridge University Press book), available at http://docs.google.com/View?docID=d2hv7k4_382f6f8zq6d (accessed Jan. 8, 2009). See also Alan Elsner, "U.S. Considering 'Odious Debt' Doctrine for Iraq," *Reuters News* (Apr. 29, 2003).

[3] See Alan Beattie, "US in Push for Iraqi Debt Relief," *Financial Times* (10 Apr. 2003) (emphasis added).

"[w]e wouldn't blame [Iraq's new] leaders if they decided that some of those [Saddam incurred] financial obligations are indeed 'odious.' And given that this is such an extreme case, international lenders probably wouldn't hold it against them for long."[4] Similarly, in a paper prepared by an economist associated with the U.S. Naval Postgraduate School in Monterey, California, it was stated that "from Iraq's point of view, the ideal solution is for creditor countries and commercial lenders to write off a substantial portion of the debt. . . . The odious debt doctrine would be an ideal way to do this. . . ."[5] In an interview with Richard Pearle, an advisor to and close associate of both Paul Wolfowitz and then–U.S. Secretary of Defense Donald Rumsfeld, the British-based *Financial Times* was reportedly told that "[i]f the French banks, the German banks and some American banks are unhappy about [debt cancellation]," then they have an opportunity to learn an important lesson concerning "lending to a vicious dictatorship."[6] Given the pull of the odious debt theory and the regularity with which it was raised in the weeks and months surrounding the initiation of the Second Gulf War, it is important to briefly examine the legal strength of the theory. After this examination, the discussion turns to international and national measures designed to prevent Iraqi oil and gas from being used to satisfy outstanding creditor claims and those provisions of Iraqi law designed to allow unilateral modification of oil and gas contracts without any appreciable legal consequence.

II. ESSENTIALS OF THE ODIOUS DEBT THEORY

It is clearly not the objective of this study to provide a sedulous examination of the intricacies and nuances of the concept of odious debt. Nevertheless, a brief description is imperative if even a cursory analysis of the theory is to be provided. The theory received virtually no attention for more than seventy-five years after its first articulation

[4] See "Iraq's Odious Debts: Don't Stick Saddam's Victims with the Bill for His Rule," *Wall St. J.* (30 Apr. 2003), available at http://odiousdebts.org/odiousdebts/index.cfm?DSP=content&ContentID=7131 (accessed Feb. 11, 2009).

[5] See Robert Looney, "Bean Counting in Baghdad: Debt, Reparations, Reconstruction, and Resources," 7 Middle East Rev. of Int'l Affairs 60, at 65–66 (Sept. 2003), available at http://meria.idc.ac.il/journal/2003/issue3/looney.pdf (accessed Feb. 11, 2009).

[6] See "Top Advisor Backs Debt Forgiveness," *Financial Times* (12 June 2003).

by Alexander Nahum Sack (Aleksandr Naumovich Zak),[7] a Russian-born émigré to Paris (and then to the United States), in 1927 in the French-language work *Le Effets de transformations des Etats sur leurs dettes publiques et autres obligations financieres* (*The Effects of the Transformation of States on Public Debts and Other Financial Obligations*). His theory states that principles of public international law cancel debts of former regimes when they are known by the creditor to have been taken on for purposes deemed unbeneficial to denizens of the debtor nation. As originally formulated, an odious debt was incurred by a despot who lacked consent of the governed, provided no benefit to the peoples ruled, and arose in a situation in which the creditor had this knowledge.[8] Numerous variations of the theory have been stated more recently, including that which allows follow-on regimes to refuse to pay debts incurred by predecessors when a creditor knew that monies provided or credit extended would be used for morally reprehensible or suspect reasons not tied to the public good.[9]

At least two significant factual situations could arise in connection with sovereign debt. The first involves the differentiation among debt incurred to a private entity, another sovereign (or an entity acting for it), and a regional or multilateral governmental financial institution. Would there be any relevance to the matter of an odious debt if the claimant were a private entity or a regional or multilateral lending institution rather than another sovereign state? The second situation involves the degree or magnitude of change that affects the outgoing government that incurred the debts of concern. In other words, what

[7] For a fascinating look at Sack's life, including his years in Russia, Estonia, Paris, and the United States (where he taught at both Northwestern University and NYU law schools), see Sarah Ludington & Mitu Gulati, "A Convenient Untruth: Fact and Fantasy on the Doctrine of Odious Debt," 43 *Va. J. Int'l L.* 595 (2008).

[8] See A. N. Sack, *Le Effets de transformations des Etats sur leurs dettes publiques et autres obligations financieres* 46–61 (1927).

[9] For some of the formulations and applications of the theory, see, e.g., Adam Feibelman, "Contract, Priority and Odious Debt," 85 *North Carolina L. Rev.* 727 (2007); Robert Howse, "The Concept of Odious Debt in Public International Law" (UNCTAD Working Paper No. 185, July 2007), available at www.unctad.org/en/docs/osgdp185_en.pdf (accessed Jan 2, 2009); Christoph G. Paulus, "'Odious Debt' vs. Debt Trap: A Realistic Help?," 31 *Brooklyn J. Int'l L.* 83 (2005); and Ashfaq Khalfan, Jeff King, & Bryan Thomas, "Advancing the Odious Debt Doctrine" (Cntr. For Int'l Sustainable Dev. Law, Paper No. COM/RES/ESJ/, 2003), available at www.odiousdebts.org/odiousdebts/publications/Advancing_the_Odious_Debt_Doctrine.pdf (accessed Jan. 22, 2009).

difference would it make that the debtor regime faced a complete territorial transfer to another or became a newly independent state, or that the debtor's experience involved the ascendancy of a radically new form of governing structure as opposed to a mere succession of administrations?

In the original formulation of the legal theory of odious debt, the difference between the degree or magnitude of governmental change and the simple passage of power from one ruling group or party to another, although not lost on Sack, did not form the basis for any distinction in legal treatment. From what is known about his formulation of the theory, the applicability of odious debt can be understood as extending only to those situations in which there existed a formal state succession – that is, one state coming into sovereign control of territory (through relinquishment or establishment of a new state or states) not previously governed by the incoming sovereign. The theory did not to extend to situations in which there was a mere passage of governmental authority from one party to another. Although not entirely clear on this point, that same view may have encompassed situations – through peaceful or other means – in which the passage of authority resulted in the state falling under the control of a radically different form of governmental philosophy.[10]

This view is certainly consistent with Sack's suggestion in his earlier 1923 work in Russian entitled *Razverstka Gosudarstvennykh Dolgor* (*Restructuring Sovereign Debt*) that the new government of the Bolsheviks/Soviets was required to honor the Tsarist government's debts to foreign creditors (although not specifically stated, no matter how odious).[11] Others have observed that Sack's 1927 work can be read as regarding the Soviet repudiation of Tsarist debts as "illegal"[12] because Sack fails to mention it in the discussion of odious debts[13] and "never [in its entirety] clearly states his views on the Soviet repudiation."[14] With Iraqi provincial elections completed by January 2009 and national elections scheduled for early 2010, this means that if Prime Minister al-Maliki's Shiite-dominated secular government were

[10] See Sarah Ludington & Mitu Gulati, supra note 7 at 597 and 617–18.
[11] See id. at 613–14.
[12] See id. at 617.
[13] See id. at 618.
[14] See id. at 617.

ousted from power by Saddarist sectarian hardliners, or even some fringe Islamist group under the control of the seriously weakened al-Qaida, Iraq would continue to be liable for payment of debts incurred by the al-Maliki government. In the event that civil society in Iraq were to collapse and the country were split apart into three separate nation-states – Kurdistan in the north, Sunnistan in the west-central, and Shiitestan in the south – or integrated by thirds into Turkey, Saudi Arabia, and Iran, the new sovereign entities governing the territory would divide Iraq's existing debts except for those deemed odious.

Invoking the theory of odious debt to relieve a mere successor government of the same preexisting nation-state of so-called odious debts of the regime just succeeded applies the theory well beyond what was originally contemplated. Since ruling parties change regularly, this dampens the interest of potential international lenders because they would never know when a new administration might raise the prospect that debts of its predecessor were considered odious. Even if the applicability of the theory were confined to those situations in which the nature or magnitude of the change in government must be regarded as so radical as to be tantamount to governance by a new sovereign, it would not allay lender apprehension in view of the practical difficulty of never being certain what sort of change would be regarded as sufficient. Extending the theory of odious debt to situations in which a significant change in governing regimes has been effected – while admirable in relieving nascent governments of some of the debts of predecessors – appears to reach beyond what was contemplated by the theory's principal architect. The long-held view is that no form of governmental change within a country changes the nature of its relations with others. As the distinguished international scholar John Bassett Moore stated decades ago: "A monarchy may be transformed into a republic, or a republic into a monarchy; absolute principles may be substituted for constitutional, or the reverse; but, though the government changes, the *nation* remains, with rights and obligations unimpaired."[15]

In the context of Iraq, modern debt-relief advocates make suggestions like "[t]he doctrine of odious debt states that when creditors lend to a dictatorial regime which they know is not using the loans to benefit

[15] See J. B. Moore, Vol. 1 *Digest of International Law* 249 (emphasis added).

the population, then debt payment cannot be demanded of those people once they are free. Iraq, which was an increasingly prosperous country until Saddam came to power, expanded the military and invaded Iran, offers a remarkably clear case where the doctrine should be applied"[16] and that "the odious debt doctrine maintains that societies should not be required to pay back a debt if the borrowed funds were not used to the benefit of the populations or were used contrary to the interests of the nation. . . . But to declare a debt odious certain conditions must be met: 1. The debt must have been incurred without the consent of the people; 2. The debt cannot have benefited the people of the debtor country; and 3. The lenders must have been aware of the preceding conditions. In the case of Iraq all three conditions are met."[17] It would certainly appear from the permitted reach of the theory's original formulation that modern advocates have attempted to extend it well beyond situations in which it was deemed by its first articulator to actually apply.

III. EARLIER APPLICATIONS OF THE THEORY

As previously intimated,[18] because it did not involve a classic situation of state succession, the originator of the theory of odious debt may have had serious doubt about whether its invocation in the context of Bolshevik/Soviet repudiation of Tsarist debts was appropriate. Another key instance of that theory's application was the U.S. repudiation after the Spanish–American War in 1898 of Spain's objectionable debts said to have been incurred relative to Cuba. The Americans argued that the debts, generated from Cuba's issuance of bonds on the international market, were "a mass of Spanish obligations and charges . . . the burdens of which, imposed on the people of Cuba without their consent and by force of arms, was one of the principal wrongs for the termination of which the struggles for Cuban independence were taken."[19]

[16] See Justin Alexander, "Downsizing Saddam's Odious Debt," Middle East Report (Mar. 2, 2004), available at www.globalpolicy.org/security/issues/iraq/contract/2004/0302odiousdebt.htm (accessed Feb. 12, 2009).

[17] See Abbas Alnasrawi, "Iraq's Odious Debt: Where Do We Go from Here?" 47 Middle East Economic Survey No. 13 (29 Mar. 2004), available at www.mees.com/postedarticles/oped/a47n13d01.htm (accessed Feb. 12, 2009).

[18] See text accompanying supra notes 10–14.

[19] See J. B. Moore, *International Arbitrations* 3591 (1906).

Furthermore, they also insisted that the debts were "created by the Government of Spain, for its own purposes and through its own agents, in whose creation Cuba had no voice."[20] Spain's loss of Cuba to the United States during that war made the United States particularly keen about helping Cuba shake off the liabilities accompanying the bonds. Antedating Sack's formal statement of the theory by more than two decades, it is understandable that no explicit reference was made to the financial obligations considered subject to repudiation because of a legal rule of odious debt.

The 1923 Tinoco arbitration between Great Britain and Costa Rica, over which U.S. Supreme Court Chief Justice Taft presided, is another application of the theory of odious debt. That dispute involved loans extended by a British financial institution to the Costa Rican government – then under the rule of the revolutionary dictator Tinoco – for what should have appeared to any reasonable person as his own use and a petroleum concession granted by the Tinoco government to a British-owned oil company then doing business in Costa Rica. Following Tinoco's forced abdication fewer than three years into his term and the restoration of the former constitution, the debt associated with the dictator's financial activities and the so-called Amory concession were legislatively nullified. In upholding the legislative nullifications, Chief Justice Taft made explicit reference to the fact that the dealings and documentary materials surrounding the debt suggested "the whole transaction . . . was full of irregularities," of which the bank must have been suspicious. The arbitration is significant for two reasons. First, as with the Spanish–American War Cuban debt negotiations, the arbitration preceded the formal articulation of the odious debt theory. By favoring the repudiation of the Costa Rican debt, it voiced the notion that debts benefiting leaders of a nation, not the people they govern, do not pass to successor regimes. Second and most significant, the arbitration involved debt repudiation in the context of governmental succession, not only state succession.

Before examining other applications of odious debt, a couple of very important points must be made regarding the examples of the Spanish–American War and the Tinoco arbitration. Concerning the latter, it is not entirely clear that the theory of odious debt served as the determinative basis for Taft's decision in favor of Costa Rica's

20 Id.

legislative repudiation of Tinoco's debts and the Amory concession. His award in the case made clear that the concession could be nullified appropriately because there had been no constitutional basis for its being granted. In other words, the demand that the new government of Costa Rica honor the contractual commitment struck by the head of state of the former government proved meaningless when it could be shown that the resulting actions were *ultra vires* the terms of the Costa Rican constitution. As for the loan debt, while it is undeniable that Taft alluded to the "irregularities"[21] surrounding the transactions, he also indicated that the new government had secured a mortgage against Tinoco's land in the country in order to guarantee repayment of any monies that formed the basis of the British bank's claim against Costa Rica. The fact that the Chief Justice referenced that his award upholding Costa Rica's repudiation of the debt "will work no injury of which Great Britain may complain, *if Costa Rica assigns all her interests in the mortgage . . . upon Tinoco's estate*"[22] certainly suggests that far more than the odious "irregularities" of the situation influenced his ultimate decision.

Concerning the Spanish–American War negotiations over the Cuban debt, it may be that the role of odiousness is somewhat clearer. Nevertheless, as the theory of odious debt put forward by modern advocates would have it, not only successor states but also successor governments would be entitled to cancel all debts considered objectionable under the basic elements of the theory. Again, there is reason to believe that as originally formulated, the theory did not go that far; nothing in the Paris Peace Conference negotiations of 1898, which aimed to resolve the Cuban debt question, suggested anything to the contrary. Even reading the positions of the American negotiators in the most favorable light regarding the odious debt theory, the fact that the situation involved territory over which Spain had formerly exercised sovereignty and was now being transferred to the United States cannot be ignored. This was not a case of mere governmental succession; this was a classic case of state succession – a case in which even Alexander Sack would have had no difficulty with suggesting that perhaps, if the factual circumstances warranted such, the

[21] See *Tinoco Arbitration*, available at www.gwu.edu/~jaysmith/Tinoco.html (accessed Feb. 13, 2009) (emphasis added).
[22] Id.

odiousness of Spanish debts would have entitled the United States to claim repudiation.

The terms of the Versailles Treaty of 1919 that addressed debt related to Poland and the efforts related to collection from the communist government in the Peoples Republic of China (PRC) on bonds issued by the Manchu Imperial Chinese Government in 1911 are other possible examples involving the odious debt theory. Article 255 (2) of the Versailles Treaty specifically exempted the new state of Poland from bearing any portion of the debt incurred when that territory undertook the German colonization program imposed by Prussian overlords.[23] Again, although it may be that debt associated with such activity can be accepted as odious due to its objectionable character, it still cannot be ignored that the situation in which it arose involved state succession, not merely governmental succession. In the case of debt taken on by the substantially earlier Manchu Imperial Chinese Government, an *aide memoire* of the PRC stated that the bonds were "one of the means" by which the issuing government "in collusion with the imperialist powers who were carving out spheres of influence in China . . . bolster[ed] its reactionary rule and repress[ed] the people,"[24] thereby at least suggesting that they could be repudiated because of their odious nature. However, in the 1982 U.S. federal court action of *Jackson v. People's Republic of China*, in which the PRC was sued by those holding some of the bonds issued by the earlier Manchu Imperial Chinese Government, the court explicitly relied on the statement of John Bassett Moore to reject the notion that a change in government alone – regardless of how innovative and different the new ideology – permits the preexisting debts from being canceled.[25]

At the international level, that same position of rejecting the applicability of the odious debt theory to situations of mere governmental succession – even when fundamentally radical in nature – seemingly received confirmation in the context of the 1996 proceedings before the Iran–U.S. Claims Tribunal associated with removal of the Shah

23 See art. 255, para. 2, Treaty of Versailles (1919), available at www.firstworldwar.com/source/versailles248-263.htm (accessed Feb. 25, 2009).

24 *Aide Memoire* submitted by the Chinese Minister of Foreign Affairs, Wu Xueqian, to the U.S. Secretary of State, George Shultz, 10 Feb. 1983, reprinted in 22 *Int'l Legal Materials* 81 (1983).

25 See 550 F. Supp. 869, 872 (N.D. Ala. 1982), rev'd on other grounds, 596 F. Supp. 386 (1984).

of Iran from power, the return of Ayatollah Khomeini from exile in Paris, and the occupation of the U.S. embassy in Tehran by Islamic fundamentalists. When Iran raised the odious debt theory to argue that financial obligations owed to the United States by earlier regimes for the purchases of military equipment dating as far back as 1948 were excused by international law following the revolutionary establishment of the Islamic Republic of Iran in 1979, that contention was squarely rejected. Iran maintained that the obligations constituted "subjugation debts" and were "not transferable" to the new republic.[26] In rejecting Iran's invocation of odious debts, the Tribunal indicated that it would "limit itself to stating that the said concept belongs to the realm of law of *state* succession."[27] It continued by observing that the law of state succession, including odious debt, "does not find application to the events in Iran. The revolutionary changes in Iran fall under the heading of state continuity, not succession."[28] The Tribunal then continued: "In spite of the change in head of State and the system of government in 1979, Iran remained the same subject of international law as before the Islamic Revolution. For when a Government is removed through revolution, the State as an international person remains unchanged and the new government generally assumes all the previous international rights and obligations of the State."[29]

IV. BASIC RULES OF STATE SUCCESSION AND PREEXISTING DEBT: RESTATEMENT

Accepting for the purposes of argument that the odious debt theory has gained status as a principle of international law – a point examined more closely in the next two sections – it clearly has an extremely limited range of applicability as shown in the previous discussion. That range of applicability to instances of actual succession of the state necessarily presents the questions of the basic rules of international law regarding succession to debt and how the odiousness of such can affect or have

[26] See *The United States of America v. The Islamic Republic of Iran* (Case No. B36), Award No. 574-B36–2, 32 Iran–United States C.T.R. 162 (1996).
[27] See id. at 176, para. 54 (emphasis added).
[28] See id.
[29] See id.

consequences for succession. In terms of attempted codifications of the relevant principles, it is appropriate to comment on both the provisions of the restatement of the law of foreign relations in the United States and those of the 1978 Vienna Convention on Succession of States in Respect of Treaties and the 1983 Convention on Succession of States in Respect of State Property, Debts, and Archives. The former convention is addressed in this section and the latter in the subsequent section.

To be sure, the matter of state succession to debts and contracts of predecessors has been and remains a controversial area of international law. From the perspective of an international lawyer in the United States, the view might be argued as reflected in the relevant provisions of the restatement of the law of foreign relations. Section 209(2) of the Restatement is most pertinent regarding succession to debts and contracts. Essentially, it provides that the debts and contracts of predecessor states were taken on by them and, therefore, should remain with them.[30] This position has a certain emotive appeal, given that the predecessor state considered, negotiated, and then bound itself to the debt or contract of concern; it was not the successor state that had been involved in any way. However, section 209 (2) provides that this basic rule can be avoided by a contrary agreement between the predecessor and successor states, and even in the absence of such an agreement, it does not operate in three exceptional situations.

The first situation in which a successor is said to assume debts or contracts of a predecessor state concerns those in which the debts or contracts of relevance relate to a portion of a predecessor state that then splits off and establishes itself as a new and independent successor.[31] Presumably, if one theme underpinning section 209 is the notion that successor states should have no liability for preexisting debts or contracts to which they have no particular connection, then there is a certain appeal in providing for successor-state liability when a preexisting debt or contract of a predecessor state related to or dealt specifically with a territory that later claimed independence and existence as a separate sovereign entity. Imagine a scenario in which the KRG of Iraq was to be the particular beneficiary of certain debts

[30] See Restatement Third of the Law of Foreign Relations, sec. 209(2)(1986).
[31] See id. at sec. 209(2)(c).

or contracts approved and negotiated by the Iraqi national government. In the event that the KRG separates from the rest of the country and established itself as an independent Kurdish sovereign state, it would be understandable to look askance at KRG claims that because Baghdad, and not Erbil, had negotiated and concluded the legal instruments creating the debt or contract that it should retain the liability, although Kurdistan may have received the benefit.

The second exceptional situation provided for in section 209(2) concerns the related instance of a benefited part of a predecessor state being subsequently absorbed by a second state. In other words, imagine that rather than the KRG separating from the rest of Iraq and establishing an independent sovereign state, it was absorbed into, say, Turkey. Again, because it had been Baghdad that negotiated and concluded the commitments that gave rise to the debt or contractual obligation, Turkey normally should not be viewed as somehow succeeding to those debts or contracts. However, given that the relevant debts or contracts related to and provided benefits for Kurdistan, it is only appropriate that whether Kurdistan subsequently established itself as an independent state or became the territory of Turkey, the successor state – not Baghdad – should be deemed liable. The language of Restatement section 209(2)(a) makes it clear that this is the preferred result – specifically "where *part* of the territory of a state becomes the territory of another state, *local public debt*, and the rights and obligations of the predecessor state under *contract relating to that territory*, are transferred to the successor state."[32]

The third exceptional situation concerns the entirety of one state being absorbed into another. If liability cannot be escaped when a successor state that emerges from a situation in which part of the territory splits off from another predecessor, which earlier incurred debts or contracts related to or dealing with the portion that is now independent, then the same result should obtain if all the territory of a predecessor is absorbed by a successor state. The successor state benefits from the public debts and contracts negotiated and concluded by the predecessor; therefore, it is reasonable to find that those debts and contracts transfer to the successor state. In accordance with this view,

[32] See id. at sec. 209(2)(a)(emphasis added).

section 209(2)(b) provides that "where a state is absorbed by another state, the public debt, and rights and obligations under contracts of the absorbed state, pass to the absorbing state."[33]

At the international level, there have been efforts to codify principles governing all such situations, but – as discussed in the next section – those efforts have not been a resounding success. Prior to summarizing those efforts, it should be observed that the restatement of the law of foreign relations in the United States also touches on the matter of state succession and international agreements. The relevancy of this is apparent due to the fact that some debts owed in the context of Iraq may involve government-to-government agreements that could share characteristics of international agreements. As for the transferability of obligation in such situations under the terms of the Restatement, section 210 takes the opposite approach to that in section 209 on debt and contracts. Essentially, section 210 provides for successor states to have an entirely clean slate concerning international agreements of predecessors – whether concern is with part of a state establishing independence from the rest or part or all of a state being absorbed by another.[34]

It is interesting that neither in the text of sections 209 or 210 nor in the accompanying commentary or reporters' notes is any reference made to the theory of odious debt. Indeed, perhaps to blunt the drawing of any possible inferences from anything that may have appeared on that score in either the provisions of the 1983 Convention on Succession of States in Respect of State Property, Debts, and Archives or the records of its accompanying negotiating history, reporters' note 8 of section 209 explicitly reminds readers that the United States did not vote in favor of the Convention and that the relevant provisions of the Restatement do not in all respects follow the terms of the convention.[35] Readers may interpret from that note what they will, but it is relatively clear that the intent is to put some distance between the language of section 209 and the analogous provisions from the 1983 Convention.

[33] See id. at sec. 209(2)(b).

[34] See id. at sec. 210.

[35] See American Law Inst., Restatement (3d) of the Foreign Relations Law of the United States, sec. 209, reporters' note 8.

V. BASIC RULES OF STATE SUCCESSION AND PREEXISTING DEBT: 1983 CONVENTION

The matters of state succession regarding both international agreements and debts were addressed internationally in detail. As mentioned previously, the former is addressed in the 1978 Vienna Convention on State Succession in Respect of Treaties (hereafter Succession in Respect of Treaties),[36] which the U.S. State Department indicated is representative of customary international law[37]; the latter is addressed in the 1983 Convention on Succession of States in Respect of State Property, Debts, and Archives (hereafter Succession in Respect of State Debts).[38]

The rules of the Succession in Respect of Treaties Convention generally favor the continuation of treaty obligations and, in that sense, opt for an approach somewhat different from that articulated in the restatement of the law of foreign relations in the United States. Article 31, for example, binds any newly formed successor state that emerges from the territory of predecessor states being merged to all the treaty commitments applicable to the territory formerly covered by the commitments.[39] In like manner, article 34 appears to require continuity of treaty commitments whenever a predecessor state splits apart into separate sovereign units – with every independent unit that remains being obligated to honor the earlier undertakings of the predecessor.[40]

The more relevant 1983 Succession in Respect of State Debts Convention – which as of this writing had garnered formal ratification from only a few states, Georgia and Ukraine being the most significant – established a more nuanced approach. Article 38 provided that so-called newly independent states (e.g., those formerly under the domination of another power) entered the community of nations with a slate clean of any portion of the debts of their former ruler, unless varied by the terms of a consensual agreement transferring debt linked

[36] Available at http://untreaty.un.org/ilc/texts/instruments/english/conventions/3_2_1978.pdf (accessed Feb. 17, 2009) (hereinafter Succession in Respect of Treaties).
[37] See Geoffrey Watson, "The Law of State Succession," in *Contemporary Practice of Public International Law* 115–16 (E. Schaffer & R. J. Snyder, eds. 1997).
[38] Available at http://untreaty.un.org/ilc/texts/instruments/english/conventions/3_3_1983.pdf (accessed Feb. 17, 2009) (hereinafter Succession in Respect of Debts).
[39] See Succession in Respect of Treaties, supra note 32 at art. 31.
[40] See id. at art. 34.

to the territory of the newly independent state so as not to endanger the fundamental economic equilibrium of that state.[41] In total contrast, article 39 provided that whenever two or more states merge so as to form a single new state, the debts of the predecessor states are assumed by the successor state.[42] This is the approach of universal succession.

Situated between the two extremes of article 38 on newly independent states and article 39 on the merger of two or more states into a single new successor are the situations in which part of the territory of one state is passed to another state; part of one state separates from another and becomes its own state; and a state dissolves, thus leading to the establishment of two or more new states. These three situations are addressed in articles 37, 40, and 41, respectively. In terms of the first situation, article 37 indicates that whenever part of a state is passed to another state in the absence of an agreement, a proportionate share of the debt that can be linked to the territory passed is transferred to the successor state.[43] Regarding the second situation, article 40 opts for a similar approach, leaving the successor state arising as a result of the separation with a proportionate share of the predecessor's debt, in the absence of an agreement to the contrary.[44] Article 41 provides that in the case of a state dissolving and becoming two or more new states, the successor state assumes debts of the predecessor based on the connection of such debt to the territory constituting each of the new states.[45]

For purposes of state succession and odious debt, however, the most relevant and instrumental portion of the Succession in Respect of State Debts Convention and its negotiating history is related to article 33, one of the general opening provisions of that portion of the Convention articulating the basic rules on state succession and debts. The bottom line is that whereas the delegates at the UN Conference that resulted in producing the Convention and the members of the International Law Commission (ILC) instrumental in drafting the predecessor documents that influenced the development of the Convention both demonstrated sympathy for and endorsement of the theory, the express

[41] See Succession in Respect of Debts, supra note 34 at art. 38.
[42] See id. at art. 39.
[43] See id. at art. 37.
[44] See id. at art. 40.
[45] See id. at art. 41.

terms of the Convention's rules deliberately refrained from directly referencing the odious debt theory *expressis verbis*. Nevertheless, it is clear that although the unhindered operation of the rules set forth in articles 37–41 of the Succession in Respect of State Debts Convention would transfer debts of predecessors in specific instances, in the event that the debts were legitimately odious, applying those rules in conjunction with article 33 can be understood as preventing them from transferring to the successor state. Moreover, there is reason to believe that although endorsed, the odious debt theory was regarded as more in the nature of *lex ferenda* – developing law – and not as *lex lata* – established law. As such, even its indirect and implicit inclusion in the ILC draft convention, and its subsequent round-about incorporation in the 1983 Debts convention, reflected both the ILC's basic charge to undertake the progressive development of international law and the world community's desire to endeavor to translate such developmental efforts into codifications binding on all states party to such.

Article 33 of the Succession in Respect of State Debts Convention provides the following: "For the purposes of the articles in the present Part, 'State debt' means any financial obligation of a predecessor State arising in conformity with international law towards a State, an international organization or any other subject of international law."[46] The ILC predecessor, article 31 of the ILC draft, provided: "For purposes of the articles in the present Part, 'State debt' means any financial obligation of a State towards another State, an international organization or any other subject of international law."[47] As demonstrated, the only difference between the draft and the Convention concerns the language "arising in conformity with international law. . . . " That language crept into the Convention by way of a Syrian amendment, proposed in the deliberations of the Committee of the Whole, which is understood as designed to provide an oblique way to incorporate the odious debt theory into the Convention, at least given that not only the Syrian[48]

[46] See id. at art. 33.

[47] See ILC Draft Articles on Succession of States in Respect of State Properties, Archives and Debts with commentaries (1981), art. 31, available at www.untreaty.un.org/ilc/texts/instruments/english/commentaries/3_3_1981.pdf (accessed Feb. 16, 2009) (hereinafter ILC Draft Articles with commentaries).

[48] See Off. Records, UN Conf. on Succession of States in Respect of State Properties, Archives and Debt, A/CONF.117/C.1/SR.30, 30th mtg. of Comm. of Whole (23 Mar. 1983) at para. 2, available at http://untreaty.un.org/cod/diplomaticconferences/

but also seven other delegates (representing the Federal Republic of Germany,[49] Morocco,[50] Iran,[51] Vietnam,[52] France,[53] Portugal,[54] and the Republic of Korea[55]) discussed the language by way of explicit reference to the theory. At least one delegate indicated that he considered the theory settled international law,[56] although the French delegate, Mr. Piris, suggested odious debt "was a difficult matter which should not . . . be dealt with in the convention." As he stated, "[t]he International Law Commission had concurred in that view" by not making explicit reference to the theory in the context of its draft convention.[57]

Any fair reading of the 1983 Conference deliberations on what became article 33 must acknowledge that whereas attention focused on some reference pulling in what eight delegates referred to as the odious debt theory – with many other delegates considering the proposed Syrian language that made its way into the article either superfluous or confusing – the primary focus in the deliberations was on the distinct question of whether the language of the article covered debts that states owed to private entities. The suggestion thus being that not only was odious debt not a preoccupation of the delegates, there also was no unanimity about how it should be handled except, apparently, on the point of not explicitly referencing it in the language of article 33. In many ways, this comports with the implications that could be drawn from the ILC's previous consideration of the theory of odious debt, which acknowledged the theory's existence and was prepared to reference it even in the official commentary to the draft articles that

succ-stateprop-archivesdbt-1983/docs/english/vol1/a_conf_117_c1_SR30.pdf (accessed Feb. 17, 2009). On the adoption (41 to 17, with 6 abstentions) by the Committee of the Whole, see 33rd mtg. (24 Mar. 1983), para. 13, available at www.untreaty.un.org/cod/diplomaticconferences/succ-stateprop-archivesdbt-1983/docs/english/vol1/a_conf_117_c1_SR33.pdf (accessed Feb. 17, 2009).

49 See id. at 31st mtg. (23 Mar. 1983) at para. 32, available at www.untreaty.un.org/cod/diplomaticconferences/succ-stateprop-archivesdbt-1983/docs/english/vol1/a_conf_117_c1_SR31.pdf (accessed Feb. 17, 2009).

50 See id. at para. 45.

51 See id. at 32nd mtg. (24 Mar. 1983), para. 6, available at www.untreaty.un.org/cod/diplomaticconferences/succ-stateprop-archivesdbt-1983/docs/english/vol1/a_conf_117_c1_SR32.pdf (accessed Feb. 17, 2009).

52 See id. at para. 18.

53 See id. at para. 34.

54 See id. at para. 64.

55 See id. at para. 74.

56 See id. (comments of Mr. Cho, Republic of Korea).

57 See supra note 53.

comprised the foundation for work on the 1983 Succession in Respect of State Debts Convention, yet limited such reference to the context of debts not passed, due to actual formal agreement between states, and refrained from expressly including the odious debt theory in the terms of draft article 31 (i.e., predecessor to article 33).[58] Is it possible that this can be explained as evidence of apprehension regarding the true status of the theory in the corpus of the rules of international law?

VI. HAS A CUSTOMARY RULE DEVELOPED THAT IS CONSISTENT WITH WHAT ADVOCATES OF DEBT RELIEF ACTUALLY WANT?

Some scholars have suggested regarding the notion of odious debt that "*expediency* is the international 'principle' which determines whether a successor will or will not assume the debts of a predecessor."[59] Indeed, the authoritative 1977 ninth report of Mohammed Bedjaoui – the ILC's special rapporteur on state succession to state property, archives, and debt, and former President of the International Court of Justice (ICJ) – can be interpreted as suggesting a similar reticence about declaring the existence of a customary rule regarding odious debt to be unquestionably demonstrable.[60] In reflecting cognizance of the ILC's role in the progressive development of international law – while also attending to the current realities of the level of maturation that the concept of odious debt had attained – Bedjaoui indicated that the notion was a "theory" developed by writers on the subject, that "State practice" suggested that such debts were excluded from succession not on the basis of obligatory legal rules but "on the basis of moral principles," and that the clean-slate approach employed with respect to such debts by writers and State practice had involved "debts contracted between European States in order to make war on other European States."[61]

[58] See ILC Draft Articles with commentaries, supra note 47 at 78–79, paras. 41–43.

[59] See Theroux & Peele, "China and Sovereign Immunity: The Huguang Railway Bonds Case," 2 *China Law Reporter* 129 at 148 (1982–83) (emphasis added).

[60] See, generally, ninth report on succession of states with respect to matters other than treaties, Doc. A/CN.4/301 and Add.1, contained in II *Yearbook Int'l Law Comm'n* 45, 67–74 (1977), available at http://untreaty.un.org/ilc/publications/yearbooks/Ybkvolumes(e)ILC_1977_v2_p1_e.pdf (accessed Feb. 18, 2009).

[61] See id. at 67, para. 116.

Collectively, this led him to the judgment that the draft articles being crafted by the ILC "should include" – rather than "must include" because already part of customary international law – reference to the notion of odious debt.[62]

It is well known that there are several basic requisites for demonstrating the existence of a principle of customary international law. Whereas it may be true that complications arise in the context of applying the odious debt theory to debts owed by Iraq, given that the situation involves an instance of governmental rather than state succession (a matter that is no minor stumbling block), the requisites essential for a customary law rule canceling such debts – even in situations of genuine state succession – appear to interject obstacles insuperable to any successful invocation of the notion. First, widely applicable rules of customary international law are required to evolve from the practice of just as widespread and diverse a number of states. Yet, as the Bedjaoui report indicated, any practice evidencing that states viewed odious debts of predecessor states as canceled essentially involved European nations and the making of war on other European nations.[63] This is far from the form of generality necessary for anything more than a limited regional, customary rule to develop. Similarly, the development of custom requires the practice of states to be driven by the existence of *opinio juris sive necessitates* – that is, a sense of legal obligation. It is inadequate for a practice to exist that arises simply from political expediency or a mere feeling of the moral rightness regarding the practice. Bedjaoui's 1977 report, however, explicitly characterized the practice underpinning the concept of odious debt as developing from "moral principles."[64] Others who have studied the situations related to odious debt suggest that any practice that can be gleaned from the historical record indicates that it is motivated by mere expediency.[65]

Without state practice driven by *opinio juris*, the basic method of proving the existence of a customary international law rule cannot be satisfied. Perhaps this accounts for the fact that, because the ILC's role is to contribute to the progressive development of international

[62] See id. at 67, para. 115.
[63] See supra note 61.
[64] See id.
[65] See supra note 59.

law, the best rapporteur Bedjaoui could suggest in his ninth report was that draft articles being formulated by the Commission to govern state succession to debts, in particular, "'*should*' include" the "*theory*" of odious debt.[66] The codification efforts of the ILC would have had no other course but to include odious debt had it been regarded as an extant aspect of customary international law – the rapporteur would then have noted the essentialness of incorporating the existence of such a rule in the language of the draft articles. Those familiar with the deliberations of the Commission appreciate that explicit language had been proposed by the rapporteur for inclusion to make clear that odious debt should be available for invocation to nullify debts of a predecessor state.[67] Such language, however, never made its way into the text of the draft articles. The reason noted for that absence was related to the idea that other more specific provisions of the draft articles could effect the same result.[68] However, in view of the rapporteur's assessment of the developmental status of odious debt in international law, is it not also conceivable that representatives to the ILC were burdened with a degree of hesitancy about a variety of matters pertaining to the theory?

This certainly seems the case in the context of the follow-on 1983 UN Conference that addressed state succession in respect to debts and its deliberations on the draft articles of the ILC. The Conference, discussed previously, evidenced a degree of hesitancy about the notion of odious debt. The Syrian proposal (which was adopted, although far from unanimously) to include language in article 33 of the 1983 Succession in Respect of State Debts Convention, to the effect that the debts subject to state succession were those "in conformity with international law," was spoken of by several delegates as incorporating the notion of odious debt into the terms of the Convention.[69] Clearly, however, there was far from a groundswell of unbounded support for the Syrian language. At least the French delegate, Mr. Piris, is on record as having indicated both the complex and difficult nature of the idea of odious debt and his reluctance about including language in the Succession in Respect of State Debts Convention that would implicate

[66] See supra note 62 (emphasis added).
[67] See ninth report, supra note 60 at 74, para. 173.
[68] See ILC Draft Articles on Succession of States in Respect of State Properties, Archives and Debts, with commentaries, supra note 47 at 79, para. 43.
[69] See supra notes 48–58 and accompanying text.

such.[70] Surely, from this and the fact that the Convention has subsequently failed to receive widespread ratification, it is not unreasonable to infer that Conference hesitancy – at least, in part – was associated with concern about the stage of legal development in which the notion of odious debt was found. While it is true that legal principles articulated in an international convention can pass into the corpus of customary international law and thereby bind even those nations not states-party to the convention, for that to occur, the convention must rise to the level of more than a drafted and endorsed document without sufficient ratifications to bring it into force.

Another scholar of international law who recently investigated the idea of odious debt permitting repudiation of financial obligations arrived independently at the conclusion that the notion has not been installed in the corpus of customary international law. Although focusing specifically on the various options available for and the wisdom of implementing an odious debt doctrine to address international debtor relief, the investigation also describes the details essential to create a principle of customary international law and offers insight on whether odious debt has managed to earn such status. In noting that many, in an offhanded way, have simply asserted that the notion of odious debt is part of custom, it was observed that under the existing "status quo, it seems clear . . . the doctrine does not have any 'legal effect,' in the sense of modifying the legal relations between debtors and creditors."[71] Furthermore, "[i]f customary international law were merely a product of specific and material choices made by responsible official policymakers [as the traditional notion of custom demands], then one could comfortably conclude that the odious debt doctrine has not become part of that body of unwritten but binding international obligations."[72] Indubitably, others may lament the view that odious debt has not yet been enshrined in customary international law or argue that such assessments are deficient in their analysis of the current status of the doctrine. Nevertheless, it is safe to say that there are serious questions about the customary status of the theory of odious debt.

[70] See supra notes 56–57 and accompanying text.

[71] See Paul B. Stephan, "The Institutionalist Implications of an Odious Debt Doctrine," 70 *Law & Contemp. Problems* 213 (2007).

[72] See id. at 221.

VII. CONCLUSION

Apart from whether any or all of Iraq's Saddam-era debts would qualify for repudiation under the doctrine of odious debt, it is clear that three major complications exist in the context of its invocation. First, it only applies to debts transferred between a predecessor and a successor state. In the case of Saddam Hussein's removal from power and the collapse of his Baath-party structure, what resulted was a mere succession of governments, not states – and no matter how systemic and deep-rooted the nature of that change that then occurred, the defense of odious debt is unavailable to the successor regime. Second, odious debt has no application to debts incurred as a result of actions of post-Saddam governmental authorities. Even if one were inclined to accept the view that some changes in regimes can be tantamount to a succession of states, the transitions among the various post-Saddam governing structures in Iraq would not be of that nature. Third, serious doubts exist about whether the concept of odious debt has earned sufficient acceptance in the practice of states – driven by the requisite *opinio juris* – to rise to the venerated level of an obligatory principle of customary international law. Surely, many hope that it can be made so by simply asserting it. However, given the limited range of and particular circumstances surrounding the actual application of the notion of odious debt, the hesitancy of the ILC in the drafting process – a hesitancy continued in the UN Conference that drafted the 1983 Succession in Respect of State Debts Convention – and the lack of receptivity on the part of the community of states to fully embrace the convention after a quarter-century, it is not unreasonable to view such assertions as missing the mark.

Part Two

The $50 billion to $70 billion owed by Iraq to foreign creditors, as reduced by the late 2009 Chinese quid pro quo regarding access to Iraqi oil in return for debt relief, and any other reductions not widely reported, would not seem subject to repudiation under an international legal notion of "odious debt." As discussed in Chapter 2, it is questionable whether customary international law even endorses the idea of debt repudiation because it was assumed under odious circumstances. Were creditors to endeavor to commence legal action to recover what they deem owed of Saddam-era debts, however, they would be sure to encounter a variety of defensive claims and strategies. In the discussion in this part, an effort is made to examine the content and the developmental background of defenses and strategies grounded in resolutions of the UN Security Council that purport to insulate Iraqi oil and gas from legal action on such debts, as well as the provisions of Iraqi domestic law – national and regional – that appear to provide Iraqi governing authorities with wide latitude to vary the terms of previously negotiated oil and gas contracts while suffering no particular legal consequences for doing so.

The basic Security Council resolution now in effect that insulates Iraqi oil and gas from legal claims, resolution 1905, was adopted at the end of December 2009. By its terms, it remains effective only until the end of December 2010. As discussed herein, the circumstances necessitating the adoption of the current resolution have bedeviled Iraq since the conclusion of the Second Gulf War and Saddam Hussein's removal from power; those circumstances had earlier been met with Security Council resolutions, including resolutions 1790 and 1859, providing much the same kind of insulation as found in resolution 1905. It is hoped that the situation in Iraq soon changes so that, at some point,

this sort of insulation can be allowed to naturally lapse, although it is plainly not likely to happen soon. Exactly how much longer the permanent members of the Security Council may be prepared to extend the protection afforded Iraqi oil and gas and the revenue from their sale is a mystery too difficult to divine. However, in light of the terms of the December 2008 SOFA between the United States and Iraq, and associated Executive Orders, it seems that at least some Security Council member states are prepared to act on their own to accord protection from legal action, even in the absence of any future extension of resolution 1905.

As for the measures of Iraqi domestic law that appear to provide legal cover for national and some regional governing authorities to alter the terms of earlier oil and gas contracts without fear of legal consequence, these emerge from both provisions of the Iraqi Constitution, the proposed national oil and gas framework law, and the oil and gas law of the KRG. How those provisions emerged and their basic thrust are considered in the following chapters. The various complex legal issues presented by either of these domestic measures, or those inherent in Security Council resolution 1905 and its predecessors, the SOFA, or implementing executive pronouncements referenced previously, are analyzed in Part Three of this study. Chapters 3 and 4 focus on the evolution of these legal instruments and their basic objectives.

3 THE DEVELOPMENT AND BACKGROUND OF SECURITY COUNCIL RESOLUTIONS 1859 AND 1905

I. INTRODUCTION

The complications associated with the odious debt theory were not only considerably avoided by the debt settlement arrangements fashioned in the context of the Paris Club, IDRO negotiations, and other bilateral and multilateral debt-relief efforts but also by the UN adoption of the Security Council resolutions that address the protection of Iraqi oil and gas and proceeds from their sale from claims by creditors. As previously mentioned, these represent an effort by the community of nations to exercise its authority under article 25 of the UN Charter to alter the thrust of general international law (e.g., *ex injuria*, *postliminium*, and the laws of war) and national law dealing with debts incident to international transactions, of holding Iraq's "feet to the fire" on its debts, and of moving in the distinct direction of providing protection from legal claims. At the outset of discussing the development and evolution of these Security Council resolutions, and before the chapters of Part Three discuss the specific legal issues presented by the terms of the resolutions, it is important to point out that the resolutions do not protect against claims arising after the restoration of Iraqi independence by the transfer of power from the Coalition Provisional Authority (CPA) in late June 2004. Restated, debts that date from the Saddam era – and that period elapsing between Saddam Hussein's removal from power and the transfer of governmental sovereignty from the CPA to the Iraqi people – are subject to the insulation provided by the current Security Council resolution 1905. Those debts resulting from actions of the Iraqi government subsequent to that time, however, do not enjoy the same protection against creditor claims.

The principal Security Council resolutions relevant to protecting Iraqi oil and gas and revenues from their sale from claims by creditors date from the period immediately after the Second Gulf War. Prior to that time, although there had been some UN-permitted export sales of Iraqi oil with attendant export earnings, Iraq was largely under UN trade sanctions designed to both pressure the government of Saddam Hussein and prevent it from acquiring the financial wherewithal to reactivate its weapons of mass destruction (WMD) programs. During that period, any limited export sales of Iraqi petroleum and petroleum products were designed to produce revenues for humanitarian needs of the Iraqi people and meet the operational needs of the UN weapons-inspection body – that is, the United Nations Special Commission (UNSCOM) – and provide monies for the UNCC to distribute to parties injured in the context of the First Gulf War. Security Council resolution 986, adopted in April 1995, addressed all of those matters and contained in paragraphs 14 and 15 the seminal language that laid the groundwork for the insulation of Iraqi oil and gas and proceeds from their sale from legal claims by creditors. In particular, after the resolution created an escrow fund for holding revenues generated by exports of Iraqi petroleum and petroleum products, paragraph 14 provided that "while under Iraqi title [petroleum and petroleum products were] immune from legal proceedings and not . . . subject to any form of attachment, garnishment or execution," that member states were to assure such immunity was implemented within their own states, and that "the proceeds of the sale [of petroleum and petroleum products] are not diverted from the purposes" of the resolution.[1] Paragraph 15 followed this by declaring that the escrow fund holding export revenues enjoyed "the privileges and immunities of the United Nations."[2]

As discussed herein, several resolutions adopted in the wake of the Second Gulf War elaborated and provided content to the regime of insulation from legal claims first introduced in resolution 986. The earliest of the post–Second Gulf War Security Council resolutions addressing oil and gas exports and protection from creditor claims

[1] See Security Council Res. 986, para. 14 (14 Apr. 1995), S/RES/986 (1995), available at http://daccessdds.un.org/doc/UNDOC/GEN/N95/109/88/PDF/N9510988.pdf?OpenElement (accessed Feb. 23, 2009).

[2] See id. at para. 15.

was resolution 1483 of May 22, 2003.[3] That resolution accomplished much, including removing the prohibition on Iraqi oil exports, thus attempting to maximize revenues to serve far more than only the purposes of meeting humanitarian needs, UNCC claims, and expenses of UNSCOM, and creating a new escrow fund and oversight body to monitor oil and gas revenues. It is obvious that with the embargo removed, the need to protect Iraqi petroleum, petroleum products, and revenues from their sale from the claims of potential creditors was considerably heightened. Resolution 1483 addressed this by providing such protection until the end of 2007 and was followed less than a year later by Security Council resolution 1546 of June 8, 2004.[4] Adopted in anticipation of the transfer of sovereign governmental power from the CPA to the Iraqi people, resolution 1546 made clear that contractual obligations entered into by Iraq after the transfer of power would not be immune from legal actions instituted by claimants. It is clear that this had the effect of distinguishing between Saddam-era and CPA-period debts and those incurred subsequent to the transition to full Iraqi independence.

On November 11, 2005, the Security Council adopted resolution 1637, which provided for an extension of authority to the oil and gas escrow fund and the fund's oversight body until the end of 2007.[5] Necessitated by continuing concern about the ethnic tensions in Iraq and uncertainty regarding the ability of the Iraqis to wisely manage their own oil and gas revenues, the Security Council believed it essential to provide continuing supervision. This was followed in late December 2006 by resolution 1723, which provided for an additional extension until the end of 2007.[6] The effect of the latter resolution was to put the escrow fund and the oversight body on a parallel time schedule with the insulation from susceptibility to legal action: December 31, 2007, now

[3] See Security Council Res. 1483 (22 May 2003), S/RES/1483 (2003), available at www.iraqwatch.org/un/unscresolutions/un-scres1483–052203.pdf (accessed Dec. 5, 2006).

[4] See Security Council Res. 1546 (8 June 2004), S/RES/1546 (2004), available at http://daccessdds.un.org/doc/UNDOC/GEN/N04/381/16/PDF/N0438116.pdf?OpenElement (accessed Feb. 23, 2009).

[5] See Security Council Res. 1637 (11 Nov. 2005), S/RES/1637 (2005), available at http://daccessdds.un.org/doc/UNDOC/GEN/N05/592/77/PDF/N0559277.pdf?OpenElement (accessed Feb. 24, 2009).

[6] See Security Council Res. 1723 (28 Nov. 2006), S/RES/1723 (2006), available at http://daccessdds.un.org/doc/UNDOC/GEN/N06/632/35/PDF/N0663235.pdf?OpenElement (accessed Feb. 24, 2009).

became the operative date for the expiration of both. Resolution 1790 of December 18, 2007 – adopted in the wake of continuing concern about the fragility of Iraq's nascent democracy – provided for another extension until the end of 2008[7]; on December 22, 2008, an additional extension was authorized until the end of 2009 by the terms of Security Council resolution 1859, and that was followed on December 21, 2009 with resolution 1905 authorizing another extension until the end of 2010.[8] Given Iraq's intense need for fiscal resources, the decline in world oil and gas prices, and the likelihood that international creditors are not anxious to see Iraq face undue economic pressure associated with allowing all claims for debt to proceed unfettered to legal action, either full international insulation from such action or selected national insulation is likely to continue into the near future.

In this chapter, the details of each relevant existing UN resolution are set forth. The objective is to provide the developmental background and contextual setting against which several of the chapters in Part Three examine the relevant legal issues that bear on the matter of the Security Council's effort to insulate Iraqi petroleum, petroleum products, and revenues from their sale from claims by creditors. Differences and divergences between the various pronouncements of the Security Council on the matter of immunity from legal action are discussed herein, with a thorough analysis of the implications in Part Three. By providing a systematic picture of the progression of the Security Council actions concerning insulation of legal claims, a greater conversancy can be acquired regarding the complexities and nuances of Security Council resolutions in this area.

II. RESOLUTION 986: POST–FIRST GULF WAR RESOLUTION

This is the first of the Security Council resolutions regarding the protection of Iraqi oil and gas and proceeds from their sale from claims by

[7] See Security Council Res. 1790 (18 Dec. 2007), S/RES/1790 (2007), available at http://daccessdds.un.org/doc/UNDOC/GEN/N07/650/72/PDF/N0765072.pdf?OpenElement (accessed Feb. 24, 2009).

[8] See Security Council Res. 1859 (22 Dec. 2008), S/RES/1859, available at http://daccessdds.un.org/UNDOC/GEN/N08/666/05/PDF/N0866605.pdf?OpenElement (accessed Dec. 27, 2008); Security Council Res. 1905 (21 Dec. 2009), S/RES/1905 (2009), available at http://daccess-dds-ny.un.org/doc/UNDOC/GEN/N09/661/91/PDF/N0966191.pdf?OpenElement (accessed Jan. 13, 2010).

creditors. The resolution is best known for starting the so-called oil-for-food program, whereby revenues from limited and supervised export sales of Iraqi petroleum and petroleum products supplied monies for addressing the humanitarian needs of the Iraqi people, as well as the compensation responsibilities of the UNCC and the funding needs of UNSCOM. As previously indicated, however, paragraph 14 of the resolution provided that Iraqi petroleum and petroleum products were immune from legal proceedings of any form while under Iraqi title, that each state was required to implement this domestically, and to ensure that proceeds from the sale of petroleum and petroleum products were not diverted from the humanitarian, compensatory, and weapons-inspection objectives of the resolution.[9] Paragraph 15 followed up by providing that the escrow account into which oil and gas revenues were deposited was to enjoy all the privileges and immunities of the UN.[10] It also must be observed that although Iraq initially balked at adherence to the resolution, it eventually relented and, within a reasonable period, the terms were supplemented with a MOU between the UN and Iraq designed to ensure full implementation of resolution 986.

Two points of interest are mentioned herein related to both the MOU dated May 20, 1996, and the terms of the resolution. Concerning the MOU, for all intents and purposes, it was focused almost exclusively on the details regarding the functioning of the oil-for-food program. It provided great detail as to how limited amounts of Iraq's oil and gas wealth would be converted – under strict UN supervision – into monies that could then be made available to meet the country's desperate humanitarian requirements, UNCC compensation responsibilities, and UNSCOM operational expenses. To be sure, the MOU referenced the notion of "privileges and immunities" in paragraphs 45–48 of the text but exclusively from the perspective of addressing protections to be accorded to those participating in the operations of the oil-for-food program, not from insulating Iraqi petroleum, petroleum products, and revenues from their sale from legal action by creditors bearing claims against Iraq.

As to the specific terms of paragraphs 14 and 15 of resolution 986, it should be observed that they evolved from draft resolution S/1995/292

[9] See supra note 1.
[10] See supra note 2.

put forward by Argentina, Oman, Rwanda, the United Kingdom, and the United States.[11] The relevant part of paragraph 14 of the draft resolution read: "...petroleum and petroleum products subject to this resolution shall while under Iraqi title be immune from legal procedures and not be subject to any form of attachment, garnishment or execution, and... all States shall take any steps that may be necessary under their respective domestic legal systems to assure this protection, and to ensure that the proceeds of the sale are not diverted from the purposes laid down in this resolution."[12] Paragraph 15 read that the "escrow account... [into which revenues from the sale of petroleum and petroleum products must be deposited] enjoys the privileges and immunities of the United Nations."[13] It is clear that the language of the immediate predecessor draft resolution was cast in precisely the same manner as paragraphs 14 and 15 of the final resolution. As a consequence, no interpretative guidance concerning the meaning of resolution 986 is to be gained from comparing and contrasting its language with that of S/1995/292.

In parsing the record of the Security Council debates regarding the draft resolution, it is apparent that the various Council members expressed concern about the need to address the humanitarian crisis that existed in Iraq but to do so in a way that avoided usurping the sovereignty of Iraq or infringing on its political integrity. Although there may be legitimate debate as to whether S/1995/292 avoided these "twin evils," the members of the Council were sufficiently satisfied to accord the resolution unanimous support in a 15-to-0 vote with no abstentions. There is nothing in the record of the Council's discussion prior to or even immediately after the vote regarding the significance of paragraphs 14 and 15 on insulation of Iraqi oil and gas and revenues from their sale from legal action prosecuted by creditors holding Iraqi debt.[14] The language of the paragraphs, however, was not entirely new to the Council because it had utilized virtually the exact language fewer than four years earlier in paragraphs 5 and 6 of

[11] See Draft Resolution A/1995/292, available at http://daccessdds.un.org/doc/UNDOC/GEN/N95/109/82/PDF/N9510982.pdf?OpenElement (accessed Feb. 26, 2009).

[12] See id. at para. 14.

[13] See id. at para. 15.

[14] See the Official Records of the Security Council, mtg. 3519 (Apr. 14, 1995), available at www.undemocracy.com/securitycouncil/meeting_3519 (accessed Feb. 25, 2009).

resolution 712.[15] Resolution 712 had been part of a two-resolution effort by the Security Council in the second half of 1991 – first with resolution 706 and then followed by 712 – to maneuver Saddam Hussein into a position of allowing the UN to permit limited post–First Gulf War exports of Iraqi oil in return for exercising disbursement control over revenues generated by such, with the goal of meeting the humanitarian needs of the Iraqi people as well as the financial demands of both the UNCC and UNSCOM. Both resolution 712 and resolution 706, however, failed to secure the necessary Iraqi cooperation. Against this background, it is understandable that resolution 986 of April 1995, which addressed much the same matter, contained in paragraphs 14 and 15 language that essentially paralleled paragraphs 5 and 6 of the earlier resolution 712. They were all directed toward protecting oil and gas and revenues from their sale from claims that would impede the basic objectives of the resolutions. As with the Security Council's deliberation in April 1995 on draft resolution S/1995/292, there is nothing in the records of the Council's 3008th meeting to discuss resolution 712 that proves instructive about the meaning of either paragraph 5 or 6.[16]

III. RESOLUTION 1483: POST–SECOND GULF WAR COMPLETE LIFTING OF EMBARGO

The efficiency of the oil-for-food program proved problematic. Because of the desire of the international community to have access to Iraqi oil and gas, the extreme need of the Iraqi people for revenues to rebuild and fund essential projects in their territory, and – with the removal of Saddam Hussein from power – the elimination of the threat of Iraq's hydrocarbon wealth being used to underwrite weapons projects, the UN found it easy to move in the direction of ending the embargo on export sales of Iraqi petroleum and petroleum products after the Second Gulf War. Paragraph 12 of the resolution established

[15] See Security Council Resolution 712, available at www.iraqwatch.org/un/unscresolutions/s-res-712.htm (accessed Dec. 4, 2006).

[16] See *Repertoire of the Practice of the Security Council 1989–92*, Chpt. 8 at 674–76, available at www.un.org/Depts/dpa/repertoire/89–92/Chapter%208/MIDDLE520EAST/item%2022_Iraq-Kuwait_.pdf (accessed Feb. 26, 2009).

the so-called Development Fund for Iraq (DFI), to be held by the CBI, and declared that the DFI's operations would be subject to scrutiny by an entity labeled the International Advisory and Monitoring Board (IAMB). Its task was to assure that irregularities did not exist with respect to the handling and disbursement of revenues collected as a result of renewed Iraqi oil and gas export sales.[17] Revenues from such sales, as well as Iraqi monies transferred from escrow accounts held by the UN and others, were to be assumed by the DFI[18] and disbursements made as paragraph 14 of the resolution declared: "... to meet the humanitarian needs of the Iraqi people... the economic reconstruction and repair of Iraq's infrastructure, the continued disarmament of Iraq... the costs of Iraqi civilian administration, and... other purposes benefiting the people of Iraq."[19] Five percent of the proceeds from oil and gas sales also were earmarked for compensation claims handled by the UNCC.[20]

Of critical importance for the purposes of this study, however, is the language of paragraph 22 in resolution 1483. That language declared the Security Council's interest in a complete restructuring of Iraq's debt and set forth the Council's decision that "until December 31, 2007, unless the Council decides otherwise, petroleum, petroleum products, and natural gas originating in Iraq shall be immune, until title passes to the initial purchaser from legal proceedings against them and not be subject to any form of attachment, garnishment, or execution, and that all States shall take any steps that may be necessary under their respective domestic legal systems to assure this protection, and that proceeds and obligations arising from sales thereof, as well as the Development Fund for Iraq, shall enjoy privileges and immunities equivalent to those enjoyed by the United Nations...."[21] In a broad sense, the language of paragraph 22 continues that which the Security Council had used nearly eight years earlier in resolution 986. Some differences exist, however, and they merit discussion.

The first difference between paragraph 22 of resolution 1483 and paragraphs 14 and 15 of resolution 986 is related to natural gas. The

[17] See Security Council Resolution 1483, supra note 3 at para. 12.
[18] See id. at paras. 17 and 20.
[19] See id. at para. 14.
[20] See id. at para. 21.
[21] See id. at para. 22.

earlier formulation of the Security Council spoke only of "petroleum and petroleum products,"[22] whereas paragraph 22 references both of these and "natural gas."[23] Second, the earlier resolution – particularly paragraph 14 – used the linguistic configuration that immunity from legal proceedings existed for petroleum and petroleum products "while under Iraqi title."[24] Conversely, paragraph 22 of resolution 1483 settled on the language that immunity existed for petroleum, petroleum products, and natural gas "until title passes to the initial purchaser."[25] The third difference concerns the notion of privileges and immunities equivalent to those enjoyed by the UN. The pertinent language of the earlier resolution – paragraph 15 of resolution 986 – referenced the Iraqi funds "escrow account . . . [as] enjoy[ing] the privileges and immunities of the United Nations,"[26] whereas paragraph 22 of resolution 1483 refers instead to the "proceeds and obligations arising from sales . . . [of Iraqi petroleum, petroleum products, and natural gas], as well as the Development Fund for Iraq . . . enjoy[ing] privileges and immunities equivalent to those enjoyed by the United Nations."[27] It is apparent that the only protection accorded by resolution 986 to "proceeds" from the sale of petroleum and petroleum products was that involving member states taking necessary steps under their domestic legal systems to "ensure that the proceeds of the sale are not diverted from the purposes" of the resolution.[28] The implications of these differences, if any, are addressed in the chapters of Part Three. For present purposes, the question is whether anything appears in either the record of the Security Council's deliberation on resolution 1483 or the history of its development that might cast an informative light on the differences.

Regarding the records of the Security Council's deliberations, it is clear that at the 4761st meeting of the Council on May 22, 2003, there was no discussion in advance of draft resolution S/2003/556, which was proposed by Spain, the United Kingdom, and the United

[22] See Security Council Resolution 986, supra note 1 at para. 14.
[23] See Security Council Resolution 1483, supra note 17 at para. 22.
[24] See Security Council Resolution 986, supra note 1 at para. 14.
[25] See Security Council Resolution 1483, supra note 17 at para. 22.
[26] See Security Council Resolution 986, supra note 1 at para. 15.
[27] See Security Council Resolution 1483, supra note 17 at para. 22.
[28] See Security Council Resolution 986, supra note 1 at para. 14.

States.[29] It was adopted in a 14-to-0 vote with no abstentions (Syria did not participate, although it indicated that it would have voted in favor had the resolution been subjected to the greater consideration that Syria had urged).[30] Paragraph 22 of that draft read precisely the same as that paragraph in resolution 1483. In the context of the postadoption statements concerning the resolution, only the delegate from Mexico, Mr. Aguilar Zinser, offered any observation on the language of paragraph 22. Specifically, with respect to the sale of petroleum resources, he noted that the terms of the resolution required that "all proceeds should be strictly channeled towards Iraq's development and the improvement of its people's living conditions."[31] While too confining, in a sense, because the resolution also clearly permitted petroleum, petroleum products, and natural gas to be used to generate monies for UNCC and UNSCOM purposes, Mr. Zinser's observation can be interpreted as stressing in a roundabout way that proceeds from the sale of Iraqi oil and gas and, by extension, the oil and gas, are not to be deemed available for satisfying claims by creditors in general.

The history of resolution 1483's development and paragraph 22 is more interesting. Essentially, two drafts preceded the May 21 draft resolution S/2003/556 by Spain, the United Kingdom, and the United States, which was adopted as resolution 1483. The first draft appeared as early as May 9, 2003, and the second on May 15, 2003. As with resolution 1483, the May 9 draft lifted the embargo, established in the CBI a separate fund (i.e., the Iraqi Assistance Fund) to receive Iraqi monies transferred from escrow accounts of the UN and others and to hold proceeds from the export sales of oil and gas, and subjected that fund to oversight by something described as an international advisory board. In paragraphs 15 and 21, the draft addressed the matter of insulation from legal claims. Paragraph 15 decreed that the Iraqi Assistance Fund would enjoy "the privileges and immunities of the United Nations."[32] In long-winded and sweeping fashion,

[29] See S/2003/556 (21 May 2003), available at www.foreignpolicy.org.tr/documents/210503.pdf (accessed Feb. 26, 2009).

[30] See Official Records of the Security Council meeting 4761 at 1, S/PV.4761 22 May 2003, available at http://daccessdds.un.org/doc/UNDOC/PRO/N03/367/58/PDF/N0336758.pdf (accessed Feb. 26, 2009).

[31] See id. at 7.

[32] See "US–UK Draft Resolution on Post-War Iraq," para. 15, *N.Y. Times* (May 9, 2003), available at www.globalpolicy.org/security/issues/iraq/document/2003/0509postwardraft.htm (accessed Feb. 26, 2009).

paragraph 21 decreed that "petroleum, petroleum products and natural gas . . . and proceeds of sales thereof, shall be immune from judicial, administrative, arbitration or any other proceedings (including any prejudgment or postjudgment attachment, garnishment, or execution or other action to satisfy judgment) arising in relation to claims, of whatever kind and whenever accrued, against Iraq or any instrumentality or agents thereof . . . and that all Member States shall take any steps under their respective domestic legal systems necessary to give full effect to this paragraph."[33] The May 15 draft substituted the DFI for the Iraqi Assistance Fund, leaving the nature of its duties largely unchanged, but did not explicitly accord the DFI "privileges and immunities of the United Nations." Furthermore, paragraph 19, in contrast with the May 9 paragraph 21, indicated that until a representative government was operational and international debt-restructuring completed, "all funds of the Development Fund for Iraq, petroleum, petroleum products and natural gas . . . and proceeds of sales thereof including indebtedness incurred in connection with such sales, shall be immune from judicial, administrative, arbitration or any other proceedings (including any prejudgment or postjudgment attachment, garnishment, or execution or other action to satisfy a judgment) arising in relation to claims, of whatever kind and whenever accrued, against Iraq or any instrumentality or agents thereof . . . and all Member States shall take any steps under their respective domestic legal systems necessary to give effect to this paragraph."[34]

It is obvious that by resolution 1483 opting for language that tied Iraqi oil and gas insulation from legal claims to the matter of passage of title to a purchaser, it evidenced a certain dissatisfaction on the part of the members of the Security Council with the effect of both the May 9 and May 15 drafts providing more blanket protection. After all, nothing in either of the two drafts would have limited immunity for Iraqi oil and gas to situations in which Iraq had indisputably relinquished title to oil and gas that it had sold to a purchaser. To secure the necessary member assent for passage of the proposed resolution, the lead drafting states thus changed the language of their proposals so as to eliminate any understanding that would suggest continuing insulation

[33] See id. at para. 21.

[34] See "US–UK–Spain Revised Draft Resolution on Post-War Iraq," para. 19, *N.Y. Times* (May 15, 2003), available at www.globalpolicy.org/security/issues/iraq/document/2003/0515reviseddraft.htm (accessed Feb. 26, 2009).

effective even after title had been transferred from Iraq. Additionally, nothing in either the May 9 or May 15 draft would have terminated the insulation from legal claims at a date certain. As written, the May 9 draft envisioned the immunity from legal action lasting until the Security Council adopted another resolution ending such, and the May 15 draft spoke in terms that would have continued the protection until an effective Iraqi government had been established and preexisting debts had been restructured. Whereas the later formulation was an improvement on the earlier one, it need only be recalled that all Iraqi debts have yet to be resolved to appreciate the indeterminate nature of such an end-date formulation.

IV. RESOLUTION 1546: REMOVAL OF INSULATION ON POST-TRANSFER OF SOVEREIGNTY CONTRACTS

Security Council resolution 1546 of June 8, 2004, marked a departure from the somewhat blanket insulation from creditor claims provided by the earlier Security Council resolutions. Resolution 1546 primarily addressed the changes in the UN regime for Iraq that would result as a consequence of the CPA relinquishing full and complete governmental authority to the Iraqi people, effective the end of June 2004. Cognizant of the facts that the sectarian situation had not stabilized to the point of complete comfort regarding Iraqi ability to make wise decisions about the spending of oil and gas revenues and that the infrastructure and public-services sectors still required extensive financial attention, the Security Council was reluctant to completely cede its earlier supervision and oversight concerning Iraqi oil and gas sales – and the spending of revenues generated thereby – although it was prepared to permit Iraq the power to designate how its monies should be spent.

In this regard, the fourteenth paragraph of the resolution's preamble "*not[ed]* the importance of providing for continued disbursements ... [of oil and gas revenues from the Development Fund for Iraq] by the Interim Government of Iraq and its successors upon dissolution of the Coalition Provisional Authority."[35] Clearly, this suggested the Council envisioned the continuation of the DFI and its functions.

[35] See Security Council Resolution 1546, supra note 4 at para. 14 of Preamble.

Moreover, with respect to the question of insulation from liability from creditor claims, paragraph fourteen of the preamble also states that the Council "[r]ecogniz[es] the benefits to Iraq of the immunities and privileges enjoyed by Iraqi oil revenues and by the Development Fund for Iraq. . . ."[36] These statements plainly presaged what was to follow particularly in the substantive language of paragraphs 24 and 27 of the resolution.

Paragraph 24 addressed the matter of the extent of oversight and supervision retained by the Security Council as opposed to the degree of autonomy relinquished to Iraq to decide how its monies should be spent. The paragraph opened by providing that on the dissolution of the CPA, the DFI funds "shall be disbursed solely at the direction of the Government of Iraq. . . ."[37] Nevertheless, the DFI was to be "utilized in a transparent and equitable manner" and was to continue its earlier function of receiving deposits of "proceeds from export sales of petroleum, petroleum products, and natural gas" as had been established by resolution 1483.[38] Furthermore, paragraph 24 provided that the IAMB "shall continue its activities in monitoring the Development Fund for Iraq. . . ."[39] The combined effect was to give Iraqi authorities discretion regarding how oil and gas monies were to be spent but to continue to collect, disburse, and supervise all such monies through the DFI and IAMB mechanisms.

Regarding insulation from claims by creditors – the more relevant matter for present purposes – paragraph 27 of resolution 1546 declares the Security Council's decision that "the provisions of paragraph 22 of resolution 1483 (2003) shall continue to apply, except that the privileges and immunities provided in that paragraph shall not apply with respect to any final judgement [sic] arising out of a contractual obligation entered into by Iraq after 30 June 2004."[40] The effect is to leave in place the insulation provided Iraq by the terms of resolution 1483. Of great significance, however, is the fact that the insulation is made completely inapplicable to claims resulting from contracts entered into by Iraq after the June 30, 2004, date scheduled for the transfer of power

[36] See id.
[37] See Security Council Resolution 1546, supra note 4 at para. 24.
[38] See id.
[39] See id.
[40] See id. at para. 27.

from the CPA to the Interim Iraqi Government. Paragraph 27 thus seems to introduce an important distinction between the period during which the initiation of legal claims is not permitted and the period during which the action of Iraqi authorities said to have given rise to such claims must have occurred. In terms of the time frame during which legal claims are not permitted, resolution 1483 establishes it as until December 31, 2007, unless otherwise extended by the Security Council. As for the time frame during which the basis for the legal claim must have accrued, paragraph 27 of resolution 1546 indicates that it extends no later than the CPA's handover of power on June 30, 2004. In other words, any claim arising from action occurring *before* the latter date may not be prosecuted prior to December 31, 2007, when resolution 1483 schedules insulation from liability to expire. Claims arising after the CPA's transfer of power on June 30, 2004, are not affected by the general insulation from liability lasting until the end of 2007.

While the history related to the development of resolution 1546 may suggest compromises and adjustments on other issues, for terms on immunity from legal claims there is only one small aspect of the language of paragraph 27 of the resolution that seems to have been the object of attention. It is related to the language of paragraph 27 that references privileges and immunities not extending to "any final judgement [sic] arising out of a contractual obligation." In the May 23, 2004, U.S.–U.K. first draft of what ultimately became resolution 1546, paragraph 17 stated the Security Council's decision "that the provisions of paragraph 22 of resolution 1483 (2003) shall continue to apply, except that the privileges and immunities provided in that paragraph shall not apply with respect to *any claim arising out of an obligation* entered into by Iraq after 30 June 2004."[41] This reference to "any claim arising out of an obligation" was then changed in the second draft dated June 1, 2004. The change appeared in paragraph 22 of the second draft and applied the privileges and immunities of the earlier resolution to "any final judgment arising out of a contractual

[41] See U.S.–U.K. Draft Resolution of 23 May 2004, available at http://news.bbc.co.uk/no1/shared/bsp/hi/pdfs/24_05_04iraqdraftscr.pdf (accessed Mar. 2, 2009) (emphasis added).

obligation" entered into after the June 30 date.[42] The June 4, 2004, third draft of what became resolution 1546 contained in paragraph 25 the same language that had appeared in the second draft.[43] From the terms of paragraph 27 of resolution 1546 quoted previously,[44] this tracks exactly the formulation settled on and adopted by the Security Council.

It is clear that there is a difference between extending insulation from legal claims to Iraqi petroleum, revenues from such, and the DFI, in relation to "any claim arising out of an obligation" entered into by Iraq no later than June 30, 2004, as opposed to "any final judgement [sic] arising out of a contractual obligation" of Iraq's carrying such date. After all, is it not reasonable to interpret "any final judgement [sic] arising out of a contractual obligation" as referring to something distinct from "any claim arising out of an obligation"? At the very least, the latter language – used in the original May 23, 2004, U.S.–U.K. draft of what in the end became resolution 1546 – can be seen as referencing any sort of claim, not only one arising from contract. Conversely, the language of paragraph 27 of resolution 1546, which refers to "any final judgement [sic] arising out of a contractual obligation," seems substantially more confined.

Beyond this first point, there is a second of equal significance. Specifically, the language of paragraph 27 of the final resolution also may be susceptible to the further observation that it suggests a difference between the privileges and immunities extended to the DFI and revenues from the sale of oil and gas, and the immunities from legal proceedings extended to Iraqi petroleum, petroleum products, and natural gas. In reviewing the language of paragraph 27 of resolution 1546, it opens with an indication that "the provisions of paragraph 22 of resolution 1483 (2003) shall continue to apply" and then finishes by excepting application of the "the privileges and immunities provided in that paragraph . . . with respect to any final judgement [sic] arising out

42 See Second Draft Resolution on Iraq (June 1, 2004), available at www.globalpolicy.org/security/issues/iraq/document/2004/0601draftresolution.htm (accessed Mar. 2, 2009).

43 See Third Draft Resolution on Iraq (June 4, 2004), available at www.globalpolicy.org/security/issues/iraq/document/2004/0604draftresolution.htm (accessed Mar. 2, 2009).

44 See text accompanying supra note 40.

of a contractual obligation entered into by Iraq after 30 June 2004." It is interesting that the language of paragraph 22 of resolution 1483 also addresses two concerns: "petroleum, petroleum products, and natural gas . . . [being] immune, until title passes to the initial purchaser, from [the institution of] legal proceedings" any sooner than December 31, 2007, unless otherwise decided by the Council; and, "proceeds and obligations arising from sales . . . [of petroleum, petroleum products, and natural gas], as well as the Development Fund for Iraq, shall enjoy privileges and immunities" accorded to the UN. Could it be that resolution 1546's application of paragraph 22 of resolution 1483 means that the UN-like privileges and immunities, which are said to be lost "with respect to any final judgement [sic] arising out of a contractual obligation entered into . . . after 30 June 2004," concerns only the DFI and revenues from oil and gas sales? Is there no merit in construing resolution 1546's linkage of final judgments on post–June 2004 contracts and the loss of privileges and immunities as limited to the DFI and revenues from hydrocarbon sales? By cross-referencing paragraph 22 of resolution 1483 – which clearly distinguishes between immunity from legal action against Iraqi petroleum, petroleum products, and natural gas on the one hand and UN-like privileges and immunities for the DFI and oil and gas revenues on the other – is it unreasonable to view resolution 1546 as ruling out legal action against Iraqi oil or gas, as such, any earlier than after December 31, 2007, regardless of the fact that the claim may be based on a final judgment in a contract arising after June 30, 2004?

Examining the record of the Security Council's deliberations on S/2004/460, the final draft of what became resolution 1546 and put forward by Romania, the United States, and the United Kingdom,[45] it appears that much the same thing happened in the Council that took place earlier with respect to resolution 1483. That is, the resolution was adopted without advance discussion by a unanimous vote, this time 15 to 0 with no abstentions. The negotiations over the various draft versions of the resolution did not result in formal minutes; did not take place prior to the vote in the context of the Council's June 8, 2004, 4987th meeting; and any commentary that was offered

[45] See S/2004/460 June 8, 2004, available at www.usembassy.at/en/download/pdf/iraq_res1546.pdf (accessed Mar. 2, 2009).

came only after the resolution's adoption. Regarding such after-the-fact commentary, nothing informative was offered by the Council's representatives concerning the matter of the reference in paragraph 27 to either the continuation of the regime established by paragraph 22 of resolution 1483 or its reference to privileges and immunities being connected to "any final judgement [sic] arising out of a contractual obligation" related to a particular date. From what was stated, however, it appears that the final wording of resolution 1546 – although perhaps not the specific language of paragraph 27 in particular – emerged as a result of amending language being actively advanced by several Security Council states, including Brazil, Chile, and Spain. Of those three states, only the Spanish representative, Mr. Yáñez Barnuevo, alluded to the DFI or the IAMB in his comments – and then only regarding the resolution's permission for Iraq to direct disbursements from the DFI, although the latter would remain subject to IAMB oversight.[46] The commentary offered by Mr. Baali of Algeria[47] and Mr. Akram of Pakistan[48] was representative of other delegates' statements, in that they highlighted the fact that the resolution contained key elements requisite for its unanimous adoption: Iraqi sovereignty, coalition military protection, a move for electoral regularity, and scrutiny regarding WMD – while noting the cooperative and flexible attitude of the resolution's chief architects. Plainly absent, however, was any illuminative discussion about the development of the language contained in the final version of paragraph 27.

V. RESOLUTIONS 1790 (2007) AND 1859 (2008): EXTENSIONS

Prior to taking up the effect of Security Council resolutions 1790[49] and 1859[50] – both adopted during the closing days of their respective calendar years and both extending the insulation from legal proceedings originally articulated in resolution 1483, it requires reiterating that

[46] See Official Records of the Security Council meeting 4987, S/PV.4987, June 8, 2004, at 12, available at http://daccessdds.un.org/doc/UNDOC/PRO/N04/380/68/PDF/N0438068.pdf (accessed Mar. 3, 2009).

[47] See id. at 3–4.

[48] See id. at 4–5.

[49] See Security Council Resolution 1790, supra note 7.

[50] See Security Council Resolution 1859, supra note 8.

the Development Fund for Iraq (DFI) was envisioned as accumulating and managing Iraqi oil and gas revenues, and its activities were subject to review by the International Advisory and Monitoring Board (IAMB). Further, separate from funds held by the DFI, various other nations around the globe also held Iraqi assets that had been frozen in the context of international sanctions against Iraq for the unsavory activities of Saddam Hussein. In accordance with the terms of paragraph 23 of Security Council resolution 1483, such funds "enjoy[ed] the same privileges, immunities, and protections as provided [Iraqi oil, gas, revenues, and the DFI by the terms of] . . . paragraph 22" of that same resolution.[51] With respect to the DFI and the IAMB, it was originally anticipated that following the CPA's transfer of authority to the Iraqi Interim Government on June 30, 2004, the functions of both would terminate on December 31, 2005. By Security Council resolution 1637 of November 11, 2005[52] and resolution 1723 of November 28, 2006,[53] the duties of the DFI and the IAMB were extended, finally to end on December 31, 2007, coterminous with the ending of the insulation from legal claims provided for in paragraph 22 of resolution 1483.

It is here that the discussion of Security Council resolutions 1790 and 1859 begins. At the time of this writing in January 2010, these two resolutions represented the most immediate predecessor statements to Security Council resolution 1905 on the matter of the UN's regime in Iraq and the roles to be played by the DFI, the IAMB, and the insulation from legal claims provided for Iraqi oil and gas revenues and other assets, whether held by the DFI or in accounts frozen by member states. Regarding these matters, both resolutions basically replicate one another, serving on a back-to-back-year basis to extend the UN regime in Iraq for another one-year period. Thus, by examining the language of Security Council resolution 1859 the nature of the earlier resolution can be ascertained, which was designed to control calendar year 2008. The operative language of resolution 1859 is in paragraph 1, which appeared the year before in paragraph 3 of resolution 1790. Paragraph 1 of resolution 1859 indicates that the roles of the DFI and the IAMB are to continue until the end of 2009, and then

[51] See Security Council Resolution 1483, supra note 17 at para. 23.
[52] See Security Council Resolution 1637, supra note 5 at para. 3.
[53] See Security Council Resolution 1723, supra note 6 at para. 3.

states, in relevant part, that the Security Council "[d]ecides to extend until 31 December 2009 . . . subject to the exception provided for in paragraph 27 of resolution 1546 (2004), the provisions of paragraph 22 of resolution 1483 (2003) . . . including with respect to funds and financial assets and economic resources described in paragraph 23 of that resolution."[54] In broad brush, the language simply means that the insulation from legal action provided by resolution 1483 to Iraqi oil, gas, revenues, the DFI, and Iraqi assets held by other countries continues until the end of 2009. However, any action based on a final judgment resulting from a contract entered into with Iraq after the CPA's transfer of sovereignty in June 2004 could be maintained, as resolution 1546 earlier had indicated. Restated, when it comes to protection against legal claims, the purpose of both paragraph 3 of resolution 1790, and paragraph 1 of resolution 1859, was to provide for extensions, for one more calendar year, of the regime established by 1483 – as modified by 1546 – in order to protect Iraqi petroleum, petroleum products, natural gas, revenues for the sale of such, the DFI, and Iraqi assets held by others.

A closer examination of the previously quoted language of paragraph 1 of resolution 1859 indicates that it extends until December 31, 2009, "the provisions of paragraph 22 of resolution 1483," including those in paragraph 23 that govern the "funds and financial assets and economic resources" described therein but subjects the foregoing to "the exception provided for in paragraph 27 of resolution 1546."[55] As mentioned in the preceding section, however, paragraph 22 of resolution 1483 distinguishes among Iraqi petroleum, petroleum products, and natural gas – which are said to be "immune" from legal action until title passes to an initial purchaser – and revenues from both their sale and the DFI – which are said to enjoy the "privileges and immunities" received by the UN.[56] Paragraph 23 of that same resolution, addressing Iraqi assets held by other member states, indicates that they enjoy "the same privileges, immunities, and protections" provided for in paragraph 22.[57] With the introduction of paragraph 27 of resolution 1546, the "privileges and immunities" of paragraph 22 of resolution 1483

[54] See Security Council Resolution 1859, supra note 8 at para. 1.
[55] See id.
[56] See supra Sec. IV.
[57] See Security Council Resolution 1483, supra note 17 at para. 23.

do not apply in the context of claims based on "any final judgement [sic] arising out of a contractual obligation" entered into by Iraq after June 30, 2004.[58] With the reference in paragraph 1 of resolution 1859 to "the provisions of paragraph 22 of resolution 1483"[59] and those of paragraph 23 dealing with "funds and financial assets and economic resources" held by member states[60] – all as subjected to the resolution 1546 exception in paragraph 27 for contracts entered after the CPA's transfer of authority in June 2004 – could questions arise regarding a distinction between the protection from legal claims entitled to be invoked by Iraqi oil and gas; revenues from their sale and the DFI; and Iraqi funds, assets, and resources held in frozen accounts by member states? In view of the language in resolution 1859 that suggests an extension until 2010 of the basic protections of resolution 1483 (as modified by paragraph 27 of resolution 1546), all the while including those Iraqi funds, assets, and resources held by others, is it not reasonable to suggest that there may be no difference in the nature of the protection from legal action provided oil and gas, revenues from their sale and the DFI, and Iraqi funds held by UN member states in frozen accounts?

In the examination of the record of the Security Council's deliberations on resolutions 1859 and 1790 – where the relevant language was first made operative – nothing appears that might be considered to offer illumination. The record of the Security Council's deliberations on the Slovakian–U.K.–U.S. draft resolution S/2007/738[61] was commented on only by Mr. Shcherbak of the Russian Federation prior to its adoption as resolution 1790 by a 15-to-0 vote with no abstentions. His comments, however, were largely directed at his country's pleasure over the successful efforts to address the many challenges faced by Iraq.[62] Other than the lengthy and informative comments offered after adoption by Mr. Hamid al-Bayati of Iraq, only the U.K.[63] and U.S.[64]

58 See Security Council Resolution 1546, supra note 4 at para. 27.

59 See Security Council Resolution 1859, supra note 8 at para. 1.

60 See id.

61 See S/2007/738 18 December 2007, available at www.undemocracy.com/S-2007–738 .pdf (accessed Mar. 4, 2009).

62 See Official Records of the Security Council meeting 5808 at 3–4 (18 Dec. 2007), S/PV.5808, available at http://daccessdds.un.org/doc/UNDOC/PRO/N07/650/65/PDF/ N0765065.pdf (accessed Mar. 4, 2009).

63 See id. at 5.

64 See id. at 4–5.

delegates offered after-the-fact observations on the resolution, and they contained nothing informative. Although Mr. al-Bayati's remarks cast no particular light on the meaning of the language of paragraph 3 in resolution 1790 – the language that ultimately became part of paragraph 1 of resolution 1859 – they represent the only reference to the important role played by the DFI and the notion of insulation from legal claims provided by paragraph 22 of resolution 1483.[65]

It was the same scenario regarding the Security Council's deliberations on S/2008/805,[66] the proposed U.K.–U.S. draft that became resolution 1859, which was adopted by a unanimous 15-to-0 vote with no abstentions. As often in the past, no comments were offered on the resolution until after its adoption. Whereas not a single delegate weighed in on interpretive matters concerning the precise language of paragraph 1 of the resolution, at least four stressed the resolution's important contribution in extending the duties of the DFI and the IAMB: Mr. Zebari[67] from Iraq, Mr. Natalegawa from Indonesia,[68] Mr. Sawer from the United Kingdom,[69] and Mr. Khalilzad from the United States.[70] Mr. Zebari also referenced Iraq's debt to others and the efforts for settlement,[71] and both Mr. Sawer and Mr. Khalilzad called specific attention to the legal immunity operating in the context of the DFI and oil revenues. Mr. Sawer indicated his nation's pleasure in the resolution's continuation of the DFI and the handling of oil revenues "and the extension of the special arrangements for that Fund, given the need for particular protection" of Iraqi assets.[72] Mr. Khalilzad noted that the resolution allowed "immunities for Iraq's funds so that these funds are available for the Iraqi Government."[73] However, in no case was illumination offered on interpretative matters presented by the paragraph 1 wording.

65 See id. at 7.
66 See S/2008/805 (19 Dec. 2008), available at http://daccessdds.un.org/doc/UNDOC/GEN/N08/666/11/PDF/N0866611.pdf (accessed Mar. 4, 2009).
67 See Official Records of the Security Council meeting 6059 at 4 (22 Dec. 2008), S/PV.6059, available at http://daccessdds.un.org/doc/UNDOC/PRO/N08?665/93/PDF/N0866593.pdf (accessed Mar. 4, 2009).
68 See id. at 6.
69 See id. at 7.
70 See id.
71 See id. at 4.
72 See id. at 7.
73 See id.

VI. CONCLUSION: RESOLUTION 1905

Up until the end of December 2009, the language of paragraph 1 of Security Council resolution 1859 controlled any question regarding the nature of UN insulation from legal action provided for both Iraqi assets and the DFI. The language is perfectly clear about two matters: (1) the protections addressed do not extend to claims based on final judgments arising from contractual obligations entered into by Iraq after the Interim Iraqi Government took over from the CPA at the end of June 2004; and (2) all other protections (as well as the functions of the DFI and the IAMB) expire December 31, 2009. So far this chapter has reviewed the entire historical background beginning with the original UN protection against legal claims and continuing through Security Council resolution 986 to what exists by way of such protection under the terms of paragraph 1 of resolution 1859. Momentarily, it will also offer some observations on the recent extension through Security Council resolution 1905 for another year of the UN protections of Iraqi oil and gas, and the continuation of the functions of the DFI and the IAMB. Several other much less clear matters are inherent in the law established by resolution 1905 and its predecessors, and these comprise the basis for the chapters in Part Three of this study.

As mentioned in the preceding descriptions of resolutions 986, 1483, 1546, 1790, and 1859, three particular matters merit reiteration. The first arises from the language of paragraph 1 of resolution 1859, the other two from the fact that the language also cross-references paragraphs 27 and 22 of resolutions 1546 and 1483, respectively. More to the point, the matters can be stated as follows: (1) the possibility that the language of paragraph 1 in resolution 1859 suggests that no distinction is intended among the types of legal protection accorded to Iraqi petroleum, petroleum products, and natural gas as opposed to revenues from their sale or the DFI, or as opposed to Iraqi assets held in frozen accounts by member states; (2) the extent of the reach of paragraph 27 of resolution 1546 in excluding from the UN legal-protection claims based on final judgments resulting from contractual obligations entered into by Iraq after the June 2004 CPA handover of governmental sovereignty; and (3) the effect of the inclusion of language in paragraph 22 of resolution 1483 that links the protection

of Iraqi petroleum, petroleum products, and natural gas from legal claims to title not being transferred by Iraq to an initial purchaser, when earlier drafts of that resolution would not have so conditioned the protection.

While these three matters comprise the basis of Part Three, the content of Security Council resolution 1905, adopted December 21, 2009, by a vote of 15 to 0 with no abstentions, merits at least brief attention.[74] As indicated earlier, resolution 1905 represents the latest iteration of the Security Council's efforts to assure that relevant UN protections and functions concerning Iraqi oil and gas and revenues from their sale do not expire. In that respect, the language of paragraphs 1 and 2 of the resolution read the same as the corresponding paragraphs of Security Council resolution 1859, the only divergence being in fixing December 31, 2010, as the new end date for UN oversight and insulation from legal proceedings. Absolutely nothing was said in the official meeting of the Security Council in which the draft predecessor of resolution 1905 – submitted by Japan, Uganda, the United Kingdom and Northern Ireland, and the United States and denominated S/2009/660 – was taken up that raises the slightest suspicion of an interest in backpedaling on UN supervision for and legal protection of Iraqi oil and gas assets. It appears there is full recognition of the vital importance to that country of oil and gas in its quest for economic survival. As reported in the provisional record of the Security Council's deliberations, the Council's president, Mr. Kafando, from Burkina Faso, stated that the consideration of and voting on draft resolution S/2009/660 was to proceed "in accordance with the understanding reached in its [i.e., the Security Council's] prior consultations."[75] The implication being, as plainly confirmed by the total absence from the official record of any comment by any Security Council representative that addresses the substance of the draft resolution, that discussions in private shaped what the resolution said and how it would be understood.

The absence of any real difference between the substantive language of the relevant paragraphs of Security Council resolution 1905 and its immediate predecessor resolutions, when coupled with silence in the

[74] Security Council Res. 1905, supra note 8.

[75] See Official Records of the Security Council meeting 6249 at 2, S/PV.6249 21 Dec. 2009, available at http://daccess-dds-ny.un.org/doc/UNDOC/PRO/N09/660/36/PDF/N0966036.pdf?OpenElement (accessed Jan. 22, 2010).

record of the Council's deliberations of the draft that became resolution 1905, suggests each of the three matters referenced above as arising from the terms of paragraph 1 of Security Council resolution 1859, and its cross-references to paragraphs 27 and 22 of resolutions 1546 and 1483, respectively, remain of considerable significance. Only the soothsayer may be capable of knowing with certainty whether it will be essential to continue the protective and insulative effect of Security Council action on Iraqi oil and gas indefinitely. It would certainly seem the odds favor the extension of such over the near-term, however. As long as that continues to be the case, a variety of nettlesome questions will rear their heads and demand the attention of lawyers representing Iraq's creditors. The eventual removal of Security Council protection is likely to present a number of other independent and no less complicated matters requiring attention.

4 THE DEVELOPMENT AND BACKGROUND OF THE IRAQI MEASURES PERMITTING DEPARTURE FROM EARLIER OIL AND GAS CONTRACTUAL COMMITMENTS

I. INTRODUCTION

In addition to the terms of Security Council resolution 1905, and its predecessor, resolution 1859,[1] two Iraqi domestic legal measures may raise questions about their effect on claimants seeking recovery for debts arising from the breach of oil and gas contracts negotiated and signed by Iraqi authorities. There are several instances in which the specifics of those measures require consideration. One instance draws on the fact that of the $50 billion to $70 billion in Saddam-era debts reportedly still owed by Iraq to foreign creditors (reduced by the late 2009 debt-for-oil field agreement with the Chinese, and any others not widely trumpeted in the media), a portion may be associated with the failure of Iraq to strictly honor the precise terms of earlier oil and gas contracts. Another instance is related to the fact that resolution 1905, by incorporating paragraph 27 of resolution 1546,[2] opens up the possibility of claims against the DFI and revenues from the sale of Iraqi oil and gas whenever such claims are based on final judgments resulting from an Iraqi contractual obligation entered into after June 30, 2004. A final instance draws on the scheduled expiry at the end

[1] See discussion of Security Council Resolutions 1905 and 1859, supra Chpt. 3, Secs. V and VI.

[2] See Security Council Resolution 1546, supra note 4 at para. 27.

of 2010 of the UN-provided insulation from legal claims – barring another extension – and points out that with its termination, breaches of oil and gas development agreements would no longer benefit from immunity.

The particular Iraqi domestic legal measures are article 141 of the 2005 Iraqi Constitution and article 54 of the 2007 KRG oil and gas law (No. 22). Article 141 of the Iraqi Constitution provides that laws of the Kurdistan region enacted since 1992 shall remain effective and KRG decisions, including court decisions and "contracts, shall be considered valid unless they are amended or annulled" by competent authority in accordance with the laws of the region, as long as "they do not contradict with the Constitution."[3] Article 54, First, of the KRG oil and gas law provides that all "agreements related to Production Sharing Contracts entered into by the [KRG] prior to the entry into force of this Law" shall be subject to Regional Council review for consistency herewith and its decisions regarding such "shall be final."[4] Article 54, Second, provides that all "authorizations and memoranda of understanding related to oil and gas . . . signed by the [KRG] prior to the entry into force of this Law" are to be regarded as "null and void" unless otherwise approved by the Regional Council.[5]

It is clear that at least with respect to article 54 of the KRG oil and gas law, this language can be significant for those who entered oil and gas arrangements with the KRG prior to the entry into force of Law No. 22 in August 2007. Any arrangements considered "agreements related to Production Sharing Contracts"[6] are subject to review by the Regional Council, and those decisions are considered final. Arrangements considered "authorizations and memoranda of understanding related to oil and gas"[7] are regarded as null and void unless otherwise endorsed by the Regional Council. The import of article 141 of the Iraqi Constitution is that all KRG contracts since 1992 – even though they may concern oil and gas – are regarded as valid and enforceable unless amended or annulled in accordance with the laws of the Region, and in a way not inconsistent with other provisions of the Iraqi

[3] See Preface, supra note 14 at art. 141.
[4] See Preface, supra note 15 at art. 54, First.
[5] See Preface, supra note 15 at art. 54, Second.
[6] See supra note 4.
[7] See supra note 5.

Constitution. The implication is that any constitutionally acceptable response to preexisting oil and gas contracts of the KRG must follow all the dictates of the Iraqi Constitution, and it is not appropriate to regard as null and void, *ab initio*, any contract simply because it is styled as an "authorization"[8] or "memoranda of understanding."[9]

In this chapter, the various complexities of both article 141 of the Iraqi Constitution and article 54 of the KRG oil and gas law (No. 22) are explored from the perspective of how they impact claims by contract holders that might be affected by either provision. To be absolutely clear, the language of both articles of Iraqi law concern only contractual arrangements with the governmental authorities of the KRG. As a consequence, undertakings between foreign entities and the central government in Baghdad, for instance, would not be controlled by the terms of either provision of law. To be sure, however, it is not that such provisions have no practical applicability. The situation reported in 2008 involving the Norwegian oil and gas company Det Norske Oljeselskap (DNO) and modifications advanced by the KRG pursuant to article 54 of the oil and gas law (No. 22) in an earlier contract concerning the Dohuk, Erbil, and Tawke licensed oil and gas areas leave no doubt in that matter.[10]

II. ARTICLE 141 OF THE IRAQI CONSTITUTION

As discussed in Section III, the language of article 54 of the 2007 KRG oil and gas law (No. 22) can be understood by contrasting it with its earlier draft predecessor, article 79 of the October 2006 Draft Petroleum Act of the Kurdistan Region of Iraq. Article 141 of the Iraqi Constitution, in providing for the continued validity of contracts entered into since 1992 by the Kurdistan Region – unless amended or annulled by competent Kurdish authorities consistent with the terms of

[8] This is the term used in art. 54, Second; see supra note 5.

[9] See id.

[10] See "DNO Wins 'Decision' in Northern Iraq," *Scandinavian Oil & Gas Mag.* (Mar. 14, 2008), available at www.scandoil.com/moxie-bm2/news/spot_news/dno-wins-decision-in-northern-iraq.shtml (accessed Mar. 6, 2009); DNO International ASA-Update on Projects in Kurdistan (Mar. 14, 2008), available at www.dn.no/finans/portal/newsPage?newt_source=feed.news.ob.ALL_SHARE&newt_newsId=DNO-1201130 (accessed Mar. 6, 2009).

the Constitution – also can benefit from examination against the background of earlier drafts. Before addressing the precise terms of those draft versions, however, recall that immediately after Saddam Hussein's removal from power, the CPA's control of Iraq was governed by the 2004 Law of Administration for the Transitional Period (TAL). Articles 53–57 of the TAL provided extensive – although not absolute – self-governance authority for the Kurdish government in the Kurdistan Region.[11] Nothing in the language of those particular provisions of the TAL, however, did more than suggest – obliquely and indirectly – that earlier Kurdistan Region contracts continued to be valid and enforceable. As a consequence, it can be interpreted that the TAL left open the question of the validity and enforceability of such contracts.

From accounts offered by those involved in the process of reviewing and commenting on the drafting of the Iraqi Constitution in mid-2005, it appears that various differing texts distinctly addressing many matters were offered during July and August. The best that can be suggested with any confidence is that the language that ultimately became article 141 first appeared in what was then article 80 of a July 28, 2005, Kurdish draft version of the Constitution. The formulation of the Kurds' draft article 80 provided for constitutional affirmation of all Kurdish legislative, executive, and judicial actions taken since the functional operations of the KRG began in 1992 and for the right of the KRG to "stop the validity of" federal measures applicable to Kurdistan, whenever deemed by the Kurds to be "irrelevant to the federal government in the region."[12] The full text of the KRG draft article 80 appeared in two paragraphs, the first stating that "[t]he executive, legislative, and judicial authorities of the Iraqi Kurdistan Region are considered as legitimate as from the time of their establishment in the year 1992" and the second that "[t]he authorities in the Kurdistan Region have the right, according to their specializations, to stop the validity of the laws and resolutions (decisions) issued in Iraq, and to decide not to apply them in Kurdistan, in part or as a whole, if they were irrelevant to the federal government in the region, in

[11] See "Law of Administration for the Transitional Period" (2004), available at www.cpa-iraq.org/government/TAL.html (accessed Mar. 6, 2009).

[12] See Ashley S. Deeks & Matthew D. Burton, "Iraq's Constitution: A Drafting History," 40 *Cornell Int'l L.J.*, 1 at 79–80 (2007).

accordance to the provision of this constitution."[13] Prior to this language on the protection of earlier actions taken by governmental authorities in the Kurdistan Region, delegates participating in the constitutional drafting process had a virtually unbounded number of other more pressing issues that they wanted addressed in the negotiating drafts of the Iraqi Constitution: the role of religion in society, the basic rights of individuals, the allocation of powers among the branches of the central government, the matter of oil and gas, and the revenues from their sale. The language proposed by the Kurds for article 80 was not widely supported when it was originally put forward.

Article 150 of the August 23, 2005, draft of the Constitution again proffered language addressing the same matter as in article 80 of the July 28 draft. In general, the language of that draft read much like what now appears in article 141 of the Constitution.[14] Essentially, this means that the reference in the earlier article 80 of the July 28 draft to the Kurds' right to "stop the validity of" federal measures considered "irrelevant to the federal government in the region" had been dropped, with attention now focused on simply validating post-1992 KRG governmental actions. The full text of article 150 of the August 23, 2005, draft stated: "Legislation enacted in the region of Kurdistan since 1992 shall remain in force. Decisions issued by the government of the region of Kurdistan, including court decisions and agreements, shall be considered valid unless it is amended or annulled pursuant to the laws of the region of Kurdistan by the competent entity and does not contradict with the constitution."[15]

Presumably, the palatability of the Kurdish proposal on this front was able to gain some traction because less than one week later, the August 28, 2005, draft of the Constitution formally transmitted by the constitutional drafting committee to the members of the Iraqi national legislature included in article 137 language that presaged article 141 of the final version of the Constitution. Article 137 of the August 28 draft stated: "Laws legislated in Kurdistan since 1992 remain in effect, and decisions made by the government of the Kurdistan

[13] Translated version of article 80, provided by email to the author on Mar. 19, 2009, by Matthew D. Burton, Attorney-Advisor, Office of the Legal Advisor, U.S. Department of State. The information provided by Mr. Burton in no way reflects the position of the Department of State or the U.S. Government.

[14] See id. at 84.

[15] See supra note 13.

region – including contracts and court decisions – are effective unless they are voided or amended according to the laws of the Kurdistan region by the concerned body, as long as they are not against the Constitution."[16]

Although the language in the August 28 draft is much like that of article 141, there are points of distinction that merit comment. In particular, the draft speaks of laws since 1992 remaining in effect unless "voided or amended,"[17] whereas article 141 of the Constitution refers to them as remaining in effect unless "amended or annulled."[18] To the extent that a difference exists between the voiding of a law and the annulment of such by the terms of article 141, from a technical perspective, only the latter remains viable under the Constitution. What is notable, however, is that both formulations permitted the amendment of relevant laws, decisions, contracts, and court judgments as long as they are amended in accordance with KRG laws and consistent with the Iraqi Constitution. Additionally, the August 28 draft references the act of voidance or amendment being taken by "the concerned [Kurdistan] body,"[19] whereas article 141 requires that such be taken by "the competent [Kurdistan] entity."[20] The clear intent is to limit the authority to vary or eliminate legislation, courts decisions, or contracts to Kurdistan Region governmental entities empowered to act, not only to those merely concerned with or interested in the matter addressed therein. The effect of the latter type of approach would have been to vest an unlimited and wide-ranging number of Kurdish entities with the authority to change or eliminate legislation, court decisions, or contracts of interest rather than confining the authority to those granted such competence.

In complete context, the central essence of article 137 of the August 28 draft and article 141 of the Constitution remains the same. Legislation, decisions, court judgments, and contracts entered into by Kurdistan since 1992 remain intact and unaltered, in the absence of an authorized Kurdish entity acting to eliminate or change such in a way that is both permitted by Kurdish law and consistent with other

16 See Aug. 28, 2005, draft of Iraqi Constitution, available at www.nytimes.com/2005/08/28/international/iraqtext_new.html?pagewanted=all (accessed Mar. 6, 2009).
17 See text accompanying id.
18 See text accompanying supra note 3.
19 See text accompanying supra note 16.
20 See text accompanying supra note 3.

provisions of the Iraqi Constitution. In other words, the fact of the adoption of a new Iraqi Constitution, even though assigning superior power regarding certain matters to the central government in Baghdad, in no way undermined or compromised the validity and effectiveness of Kurdistan Region laws, court decisions, or contracts adopted since 1992 – the period of Kurdish de facto autonomy following the U.S., U.K., and French establishment of a no-fly zone north of Iraq's 36th parallel. At the same time, however, the language of both article 141 of the Iraqi Constitution and its immediate predecessor (i.e., article 137 of the August 28 draft) clearly permitted all such laws, court decisions, and contracts to be changed or ended as long as such was affected by an appropriate Kurdish entity in accordance with the limitations imposed by other provisions of the Iraqi Constitution. The fact that article 141 speaks of an appropriate Kurdish entity being the "competent entity" signifies designation by the KRG of those entities deemed "competent" to act on such matters. Indeed, as discussed in the next section, article 54 of the 2007 KRG oil and gas law (No. 22) designates the so-called Regional Council as the competent entity to act with respect to oil and gas agreements. Similarly, the fact that article 141 also states that changes in or termination of KRG laws, court decisions, and contracts must be consonant with the Iraqi Constitution signifies that other provisions of that document must be parsed for limitations on the power to change or terminate such (see Section IV).

III. ARTICLE 54 OF THE KRG OIL AND GAS LAW (NO. 22)

As previously referenced, article 54 of the KRG oil and gas law (No. 22) designates the Regional Council – comprised of the Prime Minister, Deputy Prime Minister, Minister of Natural Resources, Minister of Finance and Economics, and Planning Minister[21] – as the entity competent to exercise the article 141 constitutional power to amend or annul contracts dealing with oil or gas and entered into prior to the law's enactment in August 2007. The precise language of article 54 is divided into two paragraphs; the first provides that all "agreements related to Production Sharing Contracts entered into by the

[21] See Preface, supra note 15 at art. 4.

Regional Government prior to the entry into force of this law shall be subject to review by the Regional Council to make them consistent with the provisions of this Law. . . . The decisions of the Regional Council . . . shall be final. . . ."[22] The second paragraph follows by stating that all "authorizations and memoranda of understanding related to oil and gas . . . signed by the Regional Government prior to entry into force of this Law shall be null and void unless . . . approved by the Regional Council."[23]

It is apparent that these two paragraphs establish a contrast between "agreements related to Production Sharing Contracts," which appear to be regarded as essentially valid yet subject to Regional Council review for consistency with the terms of the oil and gas law, and "authorizations and memoranda of understanding," which are considered null and void in the absence of Regional Council confirmation and approval. Furthermore, due to the absence in the second paragraph regarding the effect of the Regional Council's confirmation and approval process, the two paragraphs differ in that when it comes to review of agreements related to PSCs, decisions by the Regional Council are considered "final,"[24] whereas those related to requests for approval of authorizations and MOUs are susceptible to being interpreted as not having the same effect. It is obvious that the language of the article 54 paragraphs also raises questions concerning the types of negotiated undertakings to be considered agreements "related to" PSCs.[25] After all, given that the second paragraph references authorizations and MOUs "related to" oil and gas,[26] does that not inform the meaning of the reference in the first paragraph to "agreements related to" PSCs?[27]

Although reference is made to the latter question in a subsequent paragraph herein, these and other problems are addressed more extensively in Part Three of this study. Nevertheless, clearly much can be gleaned from comparing the language of article 54 with that in its immediate predecessor, article 79 of the October 2006 Draft Petroleum Act of the Kurdistan Region of Iraq. In its two sections, article 79 of the

22 See Preface, supra note 15 at art. 54, First.
23 See Preface, supra note 15 at art. 54, Second.
24 See text accompanying supra note 22.
25 See text accompanying supra note 22.
26 See text accompanying supra note 23.
27 See text accompanying supra note 22.

draft law provided that "[a]ny agreement related to Petroleum Operations entered into by the Regional Government prior to the entry into force of this Act, and approved by the Minister [of Natural Resources], shall remain in force" and that the concept of "agreement" is understood as including a "contract, license, permit, memorandum of understanding, or other legal act or dealing of any sort."[28]

Unlike article 79 of the 2006 draft law, the language of article 54 of the 2007 KRG oil and gas law (No. 22) does not vest the Minister of Natural Resources with the determinative authority regarding oil and gas agreements earlier entered but rather places that power in the Regional Council, which includes the Minister. Beyond this technical procedural point, however, article 79 made no distinction between earlier negotiated arrangements considered valid yet subject to review (i.e., "agreements related to Production Sharing Contracts") and those deemed "null and void" (i.e., "authorizations and memoranda of understanding related to oil and gas") unless subsequently approved. While article 54 of the 2007 law relies on this distinction, article 79 of the 2006 draft law only referenced the validity of earlier negotiated arrangements "related to Petroleum Operations," when approved by the Minister, and conjoined with the fact that such arrangements were sufficiently broad and encompassing to "include[] a contract, license, permit, memorandum of understanding, or other legal act or dealing of any sort." By moving away from this 2006 expansive formulation, article 54 of the KRG oil and gas law (No. 22) appears to corroborate the desire of the drafters to ensure a distinction in treatment between "authorizations and memoranda of understanding related to oil and gas" and "agreements related to Production Sharing Contracts." Arrangements considered in the latter category appear to be essentially valid but subject to Regional Council review and action; those in the former category are seen as null and void in the absence of Regional Council approval.

This distinction in article 54 between initially valid "agreements related to Production Sharing Contracts" and initially null and void "authorizations and memoranda of understanding related to oil and gas"[29] reveals the essence of its meaning when examined in the context

[28] See Draft Petroleum Act of the Kurdistan Region of Iraq, 22 Oct. 2006, available at www.krg.org/pdf/Kurdistan_Act_COM_draft_22_October_2006.pdf (accessed Apr. 4, 2009).

[29] See art. 54 quoted in text accompanying supra notes 22 and 23.

of the fact that article 79 of the 2006 draft law regarded "[a]ny agreement related to Petroleum Operations" – including "a contract, license, permit, memorandum of understanding, or other legal act or dealing of any sort" – as remaining in force if approved by the Minister of Natural Resources.[30] In other words, interpretive insight is provided about which types of contractual arrangements fall within article 54, First's, reference to "agreements related to Production Sharing Contracts," as opposed to article 54, Second's, reference to "authorizations and memoranda of understanding related to oil and gas" when that language is viewed against the background of article 79 of the 2006 draft law. As mentioned previously, this implicates the meaning of the article 54 references to arrangements that are "related to" some other form of oil and gas contractual understanding.

To comment just briefly on this matter, since article 79 of the 2006 draft law subjected "any agreement related to Petroleum Operations" to its basic rule and the term *agreement* was defined as "includ[ing] a contract, license, permit, memorandum of understanding, or other legal act or dealing of any sort,"[31] it would seem to have made no distinction on the basis of the form or nature of the contractual arrangement entered into. As long as the arrangement concerned petroleum operations (including prospecting, exploration, development, production, marketing, transportation, refining, storage, selling, exporting, and constructing or installing oil or gas facilities or structures), then it did not matter how the arrangement was structured or what descriptive label was attached. All forms of earlier negotiated oil and gas agreements were subject to precisely the same standard of ministerial approval. Article 54 of the 2007 KRG oil and gas law (No. 22), conversely, draws a straightforward distinction between "agreements related to Production Sharing Contracts" and "authorizations and memoranda of understanding related to oil and gas."[32] As a result, it is not difficult to take this distinction seriously; the problem remains, however, as to how seriously it is to be taken.

It stands to reason that "authorizations and memoranda of understanding" concerning oil and gas cannot be viewed as "*agreements related to* Production Sharing Contracts,"[33] even if involving a formal

[30] See draft of art. 79 at text accompanying supra note 28.
[31] See id.
[32] See text accompanying supra notes 22 and 23.
[33] See id. (emphasis added).

arrangement associated with a standing PSC and permitting the performance of a specified task or involving a preliminary agreement that is then followed by a more official and complete production contract. After all, article 54, Second, explicitly references "authorizations and memoranda of understanding" related to oil and gas in an entirely separate and distinct subparagraph from article 54, First's, reference to "agreement related to Production Sharing Contracts." Likewise, PSCs that permit oil and gas activity, or express in memorandum form the understanding that production is to be shared between the producers and the host country, cannot be viewed as "authorizations and memoranda of understanding related to oil and gas," because PSCs are directly mentioned in the first subparagraph of article 54. Although the question remains as to whether an oil and gas arrangement *not* in the form of a PSC, authorization, or MOU (e.g., a development and production contract or an engineering, procurement, and construction contract) can ever be considered an "agreement related to" a PSC or must always be considered an authorization or MOU "related to oil and gas," the implications of how this is answered may be significant. To begin with, if such an arrangement can be viewed as an "agreement related to" a PSC, then the arrangement is deemed valid but subject to review. If it is of such a nature that it must be seen as a form of authorization or MOU "related to" oil and gas, then it is deemed null and void in the absence of otherwise being approved by the Regional Council. Additionally, because article 54, First, speaks of Regional Council reviews of "agreements related to Production Sharing Contracts" being regarded as "final," whereas no such comparable expression appears in article 54, Second, in connection with "authorizations and memoranda of understanding related to oil and gas," the category within which a suspect arrangement is situated may be seen as affecting the possibility of internal or external review.

IV. RECONCILING ARTICLE 141 OF THE CONSTITUTION WITH ARTICLE 54 OF THE KRG OIL AND GAS LAW (NO. 22)

There may be problems with understanding the precise breadth of the oil and gas arrangements that fall within the reach of the article 54 reference to agreements "related to" PSCs or authorizations and MOUs. These problems are examined in detail in Part Three of this study. It

is perfectly clear that article 54, however, aims to express the decisions of the KRG to subject agreements related to PSCs to review by the Regional Council and to nullify and void all authorizations and MOUs relating to oil and gas, unless otherwise reaffirmed by the Regional Council. As for article 141 of the Iraqi Constitution, although it may raise questions about limitations placed on the ability of the Kurdistan Region to amend or annul oil and gas contracts entered into by Kurdish authorities since 1992, it also leaves no doubt that the basic aims are to validate all such contracts – as well as legislative measures, governmental decisions, and court judgments – and to permit appropriate Kurdish authorities to undo or terminate them, as long as it is in accordance with Kurdish law and in compliance with protections established by other provisions of the Iraqi Constitution.

This portion of the present chapter focuses on how article 54 of the KRG oil and gas law and article 141 of the Iraqi Constitution interface with one another. They both touch on the matter of preexisting contracts in the oil and gas sector and therefore share an important commonality. More significant, article 141 of the Iraqi Constitution provides the basis for the August 2007 KRG enactment of article 54 of the oil and gas law (No. 22). Article 141 essentially confirms the continued validity of earlier negotiated KRG contracts and permits their amendment or annulment pursuant to laws that the KRG might adopt. Article 54 represents the legislative enactment adopted by the KRG to express its view of whether such contracts are to be amended or annulled. By the terms of article 54, First, there can be no doubt that agreements related to PSCs are to be subject to review by the Regional Council to determine their consistency with the KRG oil and gas law (No. 22). Authorizations and MOUs related to oil and gas, however, are deemed null and void unless otherwise endorsed by the Regional Council. In taking the latter approach with respect to oil and gas authorizations and MOUs, article 54, Second, reflects an exercise of the latitude granted by article 141 of the Constitution to amend or annul earlier entered contracts. By doing nothing more than subjecting all PSC-related agreements to review for consistency with the oil and gas law, article 54, First, reflects the constitutional expression that earlier negotiated contracts are to be considered valid but can be amended.

Several points of difference between article 141 of the Iraqi Constitution and article 54 of the KRG oil and gas law cannot be ignored,

however. First, article 141 makes absolutely clear that it is only legislation, governmental decisions, court judgments, and contracts of the Kurdistan Region "since 1992" that will remain in force unless amended or annulled as specifically prescribed. Article 54 of the KRG oil and gas law speaks of agreements related to PSCs and authorizations and MOUs related to oil and gas entered into "prior to the entry into force of this Law" (i.e., August 2007). In other words, article 54 leaves open the possibility that arrangements entered into prior to 1992 could be considered enforceable – at least, after subjection to review or ratification by the Regional Council. While the fact remains that a functioning KRG structure that could enter into legal arrangements depended on both the establishment of the northern no-fly zone in April 1991, and the creation of Operation Provide Comfort administered by the UN – thereby minimizing the likelihood of any relevant pre-1992 arrangement to advance as a basis for a claim – it is clear that article 54 would be susceptible to an interpretation accommodating such, despite article 141 of the Iraqi Constitution's reference to post-1992 as the instrumental date. In this respect, however, article 141 merely erects the constitutional *minimum* that the Kurdistan Region must meet; that is, it must consider all contracts since 1992 as remaining in force unless it otherwise amends or annuls them. By the language of article 54 addressing agreements related to PSCs and authorizations and MOUs related to oil and gas entered into by the regional government "prior to the entry into force of the Law," the KRG simply declares that it may be willing – after review and consideration by the Regional Council – to validate any KRG negotiated and signed arrangements that antedate 1992, if they exist.

Another difference between article 54 of the KRG oil and gas law (No. 22) and article 141 of the Iraqi Constitution is the fact that the latter indicates explicitly that all KRG contracts "shall be considered valid" unless otherwise set aside, whereas article 54, Second, provides that authorizations and MOUs related to oil and gas "shall be null and void" unless subsequently approved by the Regional Council. This appears to place the two provisions in direct conflict with one another. After all, article 141 deems KRG contracts since 1992 valid, and article 54 expressly declares that certain types of contractual arrangements – even if entered into since 1992 – are deemed null and void. Furthermore, if article 141 of the Constitution establishes a

minimum from which the KRG is not entitled to depart – as suggested herein relative to the question of legal arrangements that the KRG may have negotiated (if any) between the time that its functionality was assured in the latter half of 1991 and the beginning of 1992 – then it certainly seems that by article 54, Second, declaring authorizations and MOUs related to oil and gas null and void, it comes up far short of that required minimum.

On the face of it, this argument certainly has a basic appeal. The language of the two provisions seems to be in direct conflict, and the line of reasoning applied to the date that any relevant arrangement must have been entered into to be considered within the protections of articles 141 and 54 leaves no other option. What makes the argument unpersuasive, however, is the fact previously mentioned that the terms of the Iraqi Constitution, including article 141, establish a structure designed to address the matter of earlier negotiated KRG laws, decisions, court judgments, and contracts. That structure is designed to make clear that by establishing a new constitutional regime – a regime now fully recognizing the authenticity and authoritativeness of the KRG and the Kurdistan Region – no intention emerges of invalidating earlier KRG measures and arrangements. It is also a structure that permits the KRG to depart from the notion of basic contractual validity by action that amends or annuls contracts in a manner comporting with KRG law and proves consistent with other provisions of the Iraqi Constitution. Here, it is important to recall the precise terms of article 141: It provides that KRG laws, decisions, court judgments, and contracts enacted, adopted, or entered into since 1992 "shall be considered valid unless they are amended or annulled pursuant to the laws of the region of Kurdistan by the competent entity in the region, provided they do not contradict with the Constitution." The idea is to empower the appropriate authorities in the Kurdistan Region to depart from the basic notion of contractual validity, if done so in accordance with Kurdish law. By the August 2007 KRG adoption of the article 54 provisions of the oil and gas law (No. 22), power recognized and granted under the Constitution was exercised by appropriate authority in the Kurdistan Region to declare earlier related authorizations and MOUs no longer valid unless subsequently ratified and confirmed by the Regional Council and to declare that earlier PSC-related agreements would be subject to Regional Council review and scrutiny. Surely, if it is possible under

the terms of article 141 of the Constitution to have an authorized entity of lesser stature and rank than the KRG parliament amend or annul previously negotiated contracts, does it not seem inconceivable that the drafters of that constitutional provision would have precluded expressions of amendment, annulment, or other treatment regarding such being voiced by the parliament? Nothing in the exact words or phraseology of article 141 suggests that alterations or nullifications of previously negotiated contracts must be executed by an authorized entity operating below the level of the parliament. Given that fact, is there not a certain reasonableness to considering the language of article 54 of the KRG oil and gas law (No. 22) as the authoritative statement and decision of whether the Constitution's basic notion of the validity of earlier KRG contracts has been departed from effectively and permissibly?

The other difference between articles 141 and 54 is related to the idea of "finality" of Regional Council decisions regarding the review of earlier agreements concerning PSCs. Article 141 declares preexisting KRG contracts essentially valid but subject to modification or termination in accordance with prescribed standards. Article 54, First, indicates that previously negotiated and signed agreements related to PSCs are to be subjected to Regional Council review for consistency with the overall terms of the oil and gas law. It then provides that all such "decisions of the Regional Council in this regard *shall be final.*" No comparable language is used in regard to authorizations and MOUs related to oil and gas. The fact that article 54, First, references Regional Council decisions being considered "final," while article 141 of the Constitution indicates nothing to that effect, comports again with the idea that the latter vests the Kurdistan Region with the constitutional authority to make the determinative judgment as to how earlier negotiated PSC-related agreements are to be handled. Although article 141 provides no indication about whether Kurdistan Region decisions to amend or annul such agreements are to be regarded as "final," it establishes a structure envisioning latitude in the KRG to pronounce such if it so decides. With the adoption of article 54, First, the Kurdistan Region exercised that latitude with respect to PSC-related agreements, abstaining from invocation of "finality" regarding authorizations and MOUs related to oil and gas. In light of the fact that any Regional Council decision to withhold ratification or confirmation from previously entered authorizations or MOUs related to oil and gas

results in the irretrievable nullification or voidance of such pursuant to the language of article 54, Second, it makes a certain degree of sense that the architects of that language would have hesitated to characterize such decisions as "final." Perhaps the palatability of abstaining from the application of finality in the context of Regional Council decisions to refuse to ratify or confirm previously negotiated arrangements – although in the form of oil- and gas-related authorizations or MOUs – seemed too draconian. An entity could be completely cut out of an earlier contractual arrangement without any recourse whatsoever. Conversely, the desire to avoid incessant appeals of Regional Council decisions on the less significant matter that PSC-related agreements be changed in one way or another to ensure consistency with the terms of the KRG oil and gas law operated to persuade the article 54 drafters of the wisdom of finality in that circumstance. With such a contractual arrangement left in place – and it being subject to alteration only to secure consistency with what was required by the KRG oil and gas law (No. 22) – the risks associated with interminable challenges to alterations was perhaps seen as outweighing the harm of imposing finality.

V. CONCLUSION

In commenting about how the terms of article 141 of the 2005 Iraqi Constitution are to be reconciled with those of article 54 of the 2007 KRG oil and gas law (No. 22), nothing in particular was offered regarding the fact that the former article references the need for any amendment or annulment of an earlier contract to be consistent with other provisions of the Constitution, whereas article 54, First, addresses Regional Council review of PSC-related agreements, focusing on consistency of such agreements with other provisions of the oil and gas law. The consistency addressed by article 141 is with the other provisions of the Iraqi Constitution; the consistency addressed by article 54, First, is with other provisions of the KRG oil and gas law. Even though two differing measures of consistency are set forth in these two provisions of law, no incompatibility or conflict attaches.

The reason for this revolves around article 141 stating nothing more than that if the KRG decides to amend or annul any earlier measures or actions, it must be done so that it meets all the protections and

safeguards of the Iraqi Constitution. Article 54, in contradistinction, states simply that the KRG has decided to exercise the amendment or annulment authority provided for in article 141, and it has chosen to regard those PSC-related agreements as subject to review for adherence to the requirements of the KRG oil and gas law (No. 22). Clearly, the consistency mandated by article 54 is with the terms of the law that governs oil and gas transactions. The consistency called for by article 141, conversely, is that of the KRG amendment or annulment of earlier measures and actions, with the requirements set forth in other provisions of the Iraq Constitution that serve to protect preexisting rights (provisions examined in the context of Chapter 6). Thus, although consistency determined by a different measure is addressed, given that these two distinct legal provisions speak to entirely distinguishable matters, it is impossible to conclude that they establish incompatible or conflicting standards.

Apart from this, it is clear from the discussion in this chapter that article 141 of the Iraqi Constitution and article 54 of the KRG oil and gas law (No. 22) both weigh in on the question of the continued validity of contracts – including oil and gas contracts – negotiated, drafted, and signed after the establishment of the KRG in 1992. The two measures raise complex issues of interpretation – issues that profit from analysis in the context of their drafting background. Article 141 of the Constitution differs from article 54 of the oil and gas law most prominently in that the former addresses every form of contract and the matter of validity, while the latter takes great pains to distinguish between PSC-related agreements and authorizations and MOUs related to oil and gas. Nevertheless, both the national legal measure and the measure adopted by the regional legislature aim to give force and effect to contractual arrangements negotiated by the Kurds subsequent to the advent of the separate and autonomous KRG. Article 54 represents the Kurds' effort to breathe life into the authority provided in article 141 of the national constitution. Both measures seek to ensure private and other parties with which deals had been executed of the continued forcefulness of such, thereby short-circuiting the likelihood of litigation or other legal recourse designed to press claims for compensation related to breach of contractual commitments.

One further matter must be addressed before leaving the discussion about the background and development of Iraqi measures that permit

the departure from preexisting oil and gas commitments: article 40 of the proposed Iraqi oil and gas law. For many months, the language of relevance was that of the February 2007 draft,[34] which contained two paragraphs, the first of which obligated the KRG to review all "Exploration and Production contracts" entered into before adoption of the law.[35] The idea was to determine the consistency of earlier contracts with the terms of the federal oil and gas law; the ultimate approval about consistency was vested in a panel of independent advisors to the central government's so-called Federal Oil and Gas Council (FOGC). Pursuant to the language of the first paragraph in the 2007 draft of article 40, the decisions of the panel of independent advisors "shall be binding."[36] The second paragraph of article 40 directed a similar review of all other exploration and production contracts not within reach of the first paragraph.[37] It bears observing that the concept of exploration and production contracts referenced in article 40 is not specifically defined in the law. However, *exploration* is defined as looking for oil and gas[38] and *production* is defined as the activity of extraction and disposal of such.[39] According to the oil and gas law, it has applicability to all "Petroleum Operations,"[40] defined broadly and inclusively to mean every imaginable sort of oil and gas activity from exploration to development, production, separation and treatment, storage, transport, and so forth.[41] Given the distinction between petroleum operations, on the one hand, and exploration and production, on the other, there is little doubt that the 2007 draft of article 40 envisioned the review obligation as related to only a limited range of preexisting oil and gas contractual arrangements – those dealing with exploration and production alone. Clearly, as discussed in the context of examining article 54 of the KRG oil and gas law (No. 22), the federal directive of article 40 concerning review – a directive of that which must be reviewed – is extended by the KRG law beyond what

[34] See Republic of Iraq Draft Oil and Gas Law, 15 February 2007, available at www.iraqrevenuewatch.org/documents/oil_law_english_20070306.pdf.
[35] See id. at art. 40(A).
[36] See id.
[37] See id. at art. 40(B).
[38] See id. at art. 4(9).
[39] See id. at art. 4(6).
[40] See id. at art. 2(a).
[41] See id. at art. 2(19).

the federal directive requires: All PSC-related agreements are subjected to review resulting in possible modification, and all authorizations and MOUs related to oil and gas are deemed outright null and void.

In May 2009, a new revised draft of the federal oil and gas law publicly surfaced.[42] As for the original directive of article 40 of the 2007 draft, the new version does not adjust the relationship between Baghdad and the KRG. However, it contains one important difference from the earlier draft that is worthy of mention. Specifically, as discussed herein, the 2007 draft article 40 applied to "Exploration and Production contracts," whereas the revised version that appeared in May 2009 speaks of reviews for all "oil contracts."[43] Because the definitional provision of the revised draft law provides that "oil contracts" are those dealing with "oil operations"[44] and the latter is defined to include essentially what the 2007 draft had defined as "Petroleum Operations" – that is, exploration, development, production, separation, processing, storage, transportation, and so forth[45] – there is no doubt that the new revised draft envisions the federal oil and gas law review directive applying to considerably more forms of oil and gas contractual arrangements than ensnared by the 2007 draft.

[42] See Republic of Iraq Draft Oil and Gas Law No.__ of 2008, 16 June 2008. For lack of legislative progress on the Oil and Gas Law, see Iraq Weekly Status Report: April 14, 2010 at 6, available at www.state.gov/documents/organization/140516.pdf (accessed April 16, 2010).

[43] See id. at art. 40.

[44] See id. at art. 4(42).

[45] See id. at art. 4(21).

Part Three

In the preceding parts of this study, attention is focused on two distinct matters. First, describing the current factual situation regarding the \$50 billion to \$70 billion of potential outstanding claims against Iraq (as modified by the late 2009 Chinese debt-for-oil agreement, and any possible others not receiving extensive coverage by the new media) and assessing the argument that such claims remain beyond recovery under the theory of "odious debt." And second, summarizing the controlling legal principles affirmed and extended in Security Council resolution 1905 (December 21, 2009), and its immediate predecessors, resolutions 1859 and 1790 as well as those set forth in article 141 of the Iraqi Constitution and article 54 of the KRG oil and gas law (No. 22) – all of which aim to establish on the international and domestic Iraqi level a measure of legal insulation from claims associated with oil and gas activity in that country. Part Three builds on the preceding chapters by mining more deeply and offering an examination and analysis of several specific legal issues that will certainly emerge and prove especially nettlesome in the context of pursuing claims for debt recovery related to oil and gas activities. The previous discussion about resolutions 1905, 1859, and 1790, and articles 141 and 54 illuminates the background concerning relevant legal considerations. However, only in sedulously critiquing what those relevant legal provisions decree in the context of specific factual situations can a more complete understanding of the precise dimensions and implications of the law be approached. Testing the relevant standards against particular situations uncloaks the various legal complexities inherent in resolution 1905 and its predecessors, and articles 141 and 54.

To that end, Part Three proceeds by first continuing the discussion just undertaken in Chapter 4 by examining several of the more

significant and controversial legal issues likely to arise under the terms of both article 54 of the KRG oil and gas law (No. 22) and article 141 of the Iraqi Constitution, in the event that any claim is made in connection with an oil and gas agreement negotiated with the KRG prior to August 2007. After all, the notions of "final[ity]" of decisions by the KRG Regional Council concerning reviewed oil and gas contracts and of arrangements "related to" either PSCs, oil and gas authorizations, or MOUs – as well as other notions broached by the specific language of these provisions of Iraqi law – are far from self-defining and automatically explicatory. Also addressed in Part Three is the particular character of the assets affected by the insulation from legal claims provided by Security Council resolutions 1905, 1859, and others, the SOFA between the United States and Iraq, and the associated Executive Orders. To the extent that such is raised about Iraqi oil and gas or revenues from their sale, this may seem straightforward. But what, if anything, is the significance of the fact that such oil, gas, or their revenues may not be held in Iraqi hands and what, if anything, results from Iraqi oil, gas, or their revenues being converted into an asset of a different character?

Another affiliated complexity considered in Part Three concerns the fact that two distinct categories of debt claims exist relative to Iraqi governmental authorities. The most numerous is likely to be that category concerning claims on long-outstanding financial obligations; the other on more recent oil and gas legal undertakings that various contracting Iraqi authorities now appear willing to complete only on the basis of the strict terms initially negotiated being altered. It is clear that debt claims arising from the latter category implicate the possible existence of Iraqi constitutional limitations on the power to change existing contractual commitments. In addition to this important issue, Part Three analyzes potential differences between pursuing legal recourse against Iraqi oil and gas or revenues from their sale, in U.S. and non-U.S. legal fora, given that Security Council resolution 1905 is scheduled to expire on December 31, 2010. Yet, the United States may see fit to use its SOFA with Iraq to extend the resolution's legal protections, thereby precluding recourse to legal action within the United States. Given the possibility that recourse to such action may be had elsewhere, also examined in a separate chapter is the matter of what resolutions 1905, 1859, and 1790 – as well as the provisions of

the SOFA and the associated Executive Orders – state about seeking to satisfy, by subsequent action in U.S. courts, judgments or awards rendered in foreign jurisdictions.

The importance of each matter mentioned cannot be overemphasized; neither can the significance of these chapters be stressed too much. It is one thing to have an abstract understanding of the magnitude of possible legal claims against Iraq for the debts it reportedly owes to others as well as the fundamental legal principles stated in controlling Security Council resolutions, provisions of Iraqi national law, or other and supplementary international agreements. It is something entirely different to appreciate the nature and intricacy of the actual legal issues that confront those who may prosecute those claims and for the nuances of possible answers to each issue confronted. In endeavoring to weave these various objectives into a coherent tapestry that provides a critical portion of a larger picture regarding claims for debts owed by Iraq, perhaps a helpful goal is at least modestly served. In transitioning to the substance of Part Three, however, readers must be admonished to prepare for a detailed and hypertechnical dissection of specific words in relevant legal provisions. Especially in Chapters 5 through 8, much that is discussed involves the seemingly repetitious tedium of the same handful of international or domestic legal measures. The sole distinguishing feature from chapter to chapter being, therefore, that whereas one chapter focuses on a certain set of words used in a particular measure, another chapter is concerned with distinct words from that very same measure.

5 THE CONCEPTS OF "FINALITY," "RELATED TO," AND ASSORTED MATTERS: ARTICLE 54 OF THE KRG OIL AND GAS LAW (NO. 22) AND ARTICLE 141 OF THE IRAQI CONSTITUTION

I. INTRODUCTION

It will be recalled that article 54 of the KRG oil and gas law (No. 22) has two paragraphs, which provide that: (1) PSC-related agreements entered into by the KRG prior to the law's adoption must undergo review by the Regional Council to ensure consistency with the provisions of the new law; and (2) similarly dated authorizations and MOUs related to oil and gas are to be regarded as null and void. The first paragraph of the article also provides that Regional Council decisions about consistency of reviewed agreements are considered final.[1] It was discussed previously that article 141 of the Iraqi Constitution declares that the KRG is vested with the authority to amend or annul contracts that it had entered into since 1992, as long as it is done consistently with other provisions of the Constitution. That article also declares all such KRG contracts not amended or annulled to be valid and fully in force.[2]

Imagine an enterprise that had negotiated and successfully concluded sometime after 1992 – and most probably after Saddam Hussein's ouster in 2003, when the KRG was at much greater liberty

[1] See supra Chpt. 4, Sec. III (discussing art. 54).
[2] See supra Chpt. 4, Sec. II (discussing art. 141).

to assert a more extensive form of civic autonomy – a contractual arrangement with the KRG concerning oil and gas situated in the Kurdistan Region. If the government of that region were desirous of either subjecting that arrangement to Regional Council review and alteration to bring it in line with what is represented as the terms of the KRG oil and gas law (No. 22) or nullifying and voiding the arrangement because of its status as a mere authorization or MOU, rather than an agreement related to a PSC, such an action would certainly cause either the consternation of the affected enterprise or at least interest in ascertaining the extent to which it could invoke legal rights in the effort to push back. To be sure, numerous oil and gas arrangements of the type suggested were in fact negotiated between the KRG and overseas entities; from all indications, article 54 proceedings have occurred with respect to such. The most prominent proceeding was a widely reported instance in 2008 of an article 54 review (see Chapter 4) that involved the Norwegian oil and gas company DNO and its earlier license from the KRG concerning the Dohuk, Erbil, and Tawke oil fields.[3] Another review in the same year that did not receive as much media attention involved the Genel Enerji/Addax Petroleum PSC with the KRG related to the Taq Taq and Chirmila area deposits.[4]

It is difficult to develop a comprehensive picture of other earlier oil and gas arrangements with the KRG, much less one that details all those oil and gas arrangements entered into by the KRG between 1992 and Saddam Hussein's removal from power in spring 2003, and then does the same from 2003 to August 2007 when the oil and gas law (No. 22) was enacted. What is known, however, is that an official media release from the KRG – reprinted by the private group Intellibrief and regarding a November 4, 2007, meeting of the Regional Council, at which both new PSCs and those predating the adoption of the oil and gas law (No. 22) were considered – indicated the completion of article 54 reviews on at least five old oil and gas PSCs. Those old PSCs were with DNO and Genel Enerji/Addax Petroleum – the companies previously mentioned – and with WesternZagros, Hawler Energy/A&T

[3] See text accompanying supra Chpt. 4, note 4.

[4] See "Addax Petroleum Corp – Update on KRG Discussion" (Feb. 29, 2008), available at www.reuters.com/article/pressRelease/idUS146810±29-Feb-2008±RNS20080229 (accessed Apr. 4, 2009).

Petroleum, and the triumvirate of Shakal/Trilax/Petoil.[5] Press reports indicated that the arrangement with Shakal/Trilax/Petoil was concluded as early as January 2003,[6] while that with DNO was concluded in June 2004,[7] and the arrangements with Genel Enerji/Addax Petroleum and WesternZagros were concluded in May 2006.[8] The KRG media release about the November 4, 2007, Regional Council meeting offered nothing regarding other forms of contractual arrangements other than PSCs. Therefore, it is difficult to know whether other arrangements – and if so, how many – were then or at other times subject to possible review and modification by the Regional Council to ensure consistency with the terms of the oil and gas law (No. 22) or to leave in place the nullification and voidance of the arrangements pursuant to the directive of the second paragraph in article 54.

It is clear that potential exists for legal challenges, in court or before arbitral bodies, to any Regional Council action modifying or annulling oil and gas arrangements negotiated and successfully concluded between the KRG and foreign entities from 1992 to August 2007. In the context of challenges seeking recovery for breach of contractual commitment, ample opportunity exists for examining questions such as the significance of the article 54 reference to Regional Council decisions being considered "final."[9] Opportunity also exists for examining the meaning of that same article's references to agreements "related to" PSCs and authorizations and MOUs "related to" oil and gas,[10] as well as associated questions such as whether the article's very terms extend beyond those contractual arrangements

[5] See "Kurdish Government Signed 7 New Petroleum Contracts" (Nov. 8, 2007), available at http://intellibriefs.blogspot.com/2007/11/kurdish-government-signed-7-new-petroleum-contracts.html (accessed Apr. 5, 2009).

[6] See "Baghdad Lays Down the Law on KRG Contracts," zawya.com (2008), available at www.zawya.com/story.cfm/sidv51n01-1TSO1/Baghdad%20Lays%20Down%20Law%20on%20KRG%20Contracts (accessed Apr. 10, 2009).

[7] See "Status Northern Iraq" (DNO Press Release, Nov. 29, 2005), available at http://www.dno.no/Press-Media/Press-Releases/2005/Status-Northern-Iraq/ (accessed Apr. 10, 2009).

[8] See "Oil Firms to Submit $1bn Iraq Development Plan," arabianbusiness.com (Sept. 2, 2007), available at www.arabianbusiness.com/498915-oil-firms-to-submit-$1bn-iraq-development-plan?In=en (accessed Apr. 10, 2009) (Genel Enerji/Addax); WesternZagros: History, available at www.westernzagros.com/about/history.html (accessed Apr. 10, 2009).

[9] See supra note 1.

[10] See id.

entered into directly with the KRG and encompass third-party arrangements concluded by the KRG's foreign-entity contractors with providers of various supplies and services. This chapter examines those questions in detail, thereby endeavoring to add informative glosses on the complex interpretive difficulties presented by the language of article 54 of the KRG oil and gas law (No. 22).

Before addressing the complicated and specific questions regarding the meaning of certain words and phrases employed in article 54, it bears noting that alternatives to the more conventional and standard so-called *textualist* interpretive methodology exist and have their own enthusiastic proponents and adherents. Each methodology could have been observed as having relevance in the development and background of the various Security Council resolutions reviewed in Chapter 3, or the provisions of Iraqi law permitting departure from preexisting oil and gas contracts examined in Chapter 4. However, explicit reference to the textualist interpretive methodology and its challengers – the *intentionalist* and the *teleological* approaches – was postponed until this chapter since it is here that the most probing, substantial, detailed, and careful parsing of the language of certain provisions of the law occurs to any substantial extent.

To one schooled in international law, the textualist methodology of interpretation elevates the ordinary meaning of words used in a document that raises interpretive issues to a stature above[11] that which the intentionalist approach accords to the state of mind of the drafters (as divined from what the negotiating records and preparatory instruments reveal)[12] or that which the teleological approach accords to an effectuation of the essential or overall purpose of the document being

[11] See, e.g., the approach of article 31 in the 1969 Vienna Convention on the Law of Treaties, U.N. Doc. A/CONF.39/27 at 289 (1969), 1155 U.N.T.S. 331, reprinted in 8 *Int'l Legal Materials* 679 (1969). Article 31(1) provides: "A treaty shall be interpreted in good faith in accordance with the *ordinary meaning* to be given to the terms of the treaty in their context and in light of its object and purpose" (emphasis added). The language's prioritization of "ordinary meaning" suggests to some a clear hierarchy favoring textualism, especially in light of the fact that article 32 of the Vienna Convention on the Law of Treaties goes on to characterize the legislative or negotiating record as a "supplementary" means of interpretation.

[12] It cannot be denied that article 32 of the Vienna Convention, id., establishes legislative or negotiating records as having a mere "supplementary" status. Some have indicated that this results from the fact that many sign on to international agreements long after the original negotiations in which one would have had an opportunity to express its

construed.[13] Although international and national law cases exist in which the textualist approach is relied on,[14] it is not uncommon for a practitioner or scholar of international law to have recourse to all three interpretive methodologies.[15] Due weight is accorded initially to the precise wording of the document or legal rule being disputed, but it is given in the context in which the relevant wording appears and in light of the fundamental object and purpose sought by the document or rule in which it is contained. In the event that these sources still leave unanswered certain questions regarding the meaning of the pertinent language or, perhaps, in the context of attempting to fully comprehend the basic object and purpose of the overall document or rule itself, practitioners and scholars may resort to the *travaux préparatoires*, or negotiating record, of relevance. The following analysis of the language of article 54 of the KRG oil and gas law (No. 22) relies heavily on the textualist methodology, with emphasis on interpreting language in light of the total context in which it is used. Infrequently, however, there is recourse to arguments that resemble those of the teleologist. To the extent that analysis invokes earlier KRG legislative efforts, such as the October 2006 Draft Petroleum Act of the Kurdistan Region of Iraq,[16]

intent concerning the meaning of subsequently controversial provisions. Given the consequent disadvantage resulting therefrom, it would seem appropriate to place presumed or inferred intent below ordinary meaning on the interpretive hierarchy. See David J. Bederman, International Law Frameworks at 35 (2001).

13 See the fact that article 31(1) of the Vienna Convention, supra note 9, plainly references ascertaining the "ordinary meaning" of words not only in their context but also "in light of its [i.e., the international agreement's] object and purpose."

14 On the international level, see, e.g., the 1932 Advisory Opinion of the Permanent Court of International Justice in *Interpretation of the 1919 Convention Concerning Employment of Women During the Night*, [1932] P.C.I.J., Ser. A/B, No. 50 (interpreting art. 3 of the 1919 Convention and deciding that it was obligated to follow the strict text of the article, despite the fact that doing so might be seen as a less-than-progressive result by barring women from working in managerial or supervisorial positions if the work involved night-time hours). On the national level, American lawyers are intimately familiar with the so-called plain-meaning rule and, in foreign jurisdictions, comparable approaches can be found that stress textualism. See, e.g., the 1958 decision of the Austrian Constitutional Court in *Case on the Interpretation of the Austrian State Treaty* (Aust. Const. Ct. 1958), reprinted in 86 *J. du Droit International* 835 (1959) (interpreting the term *airplane* used in a post–World War II international agreement by referencing the strict language of other texts used in the same agreement and thereby concluding that the term covered both powered and gliding aircraft).

15 For a decision in which the ICJ was receptive to the teleological approach, yet cautioned against too far an extension of it, see "Interpretation of Peace Treaties with Bulgaria, Hungary and Romania," *Opinion* [1950] ICJ 65, 221.

16 See infra note 32.

the analysis also shares in the intentionalist methodology to assist in this interpretive exercise.

II. THE SIGNIFICANCE OF THE CONCEPT OF "FINAL[ITY]"

Before offering any meaningful comments regarding the reference in article 54, First, to the fact that decisions of the Regional Council concerning agreements related to PSCs are considered "final,"[17] recall the two points made in Chapter 4. First, article 54 operates in conjunction with article 141 of the Iraqi Constitution – the latter confirming the continued validity, among other things, of all KRG contracts entered into since 1992 unless treated otherwise under KRG law, and in a way not inconsistent with the terms of the Iraqi Constitution.[18] Second, article 54 represents the precise way by which the KRG chose to reflect its desires regarding treatment of all oil and gas contracts entered into between 1992 and August 2007. The Regional Council referenced in article 54 is a body comprised of the Prime Minister, Deputy Prime Minister, and the Ministers of Natural Resources, Finance, and Planning,[19] all knowledgeable about or with access to experts who have the requisite knowledge about contractual arrangements concerning oil and gas.

With this background, three principal observations are warranted about the fact that article 54, First, invokes the concept of "final[ity]," a concept that seems to suggest that once Regional Council decisions regarding preexisting agreements have been made, no recourse exists for reconsidering, reevaluating, or challenging those decisions. One of the principal observations addresses the fact that article 54, First, calls on finality with respect to PSC-related agreements, whereas article 54, Second, voices no similar call in the context of Regional Council consideration of authorizations and MOUs related to oil and gas. Another observation concerns the interface between the article 54 invocation of finality and the article 141 declaration in the Iraqi

[17] See supra note 1.
[18] See supra note 2.
[19] See art. 4, Oil and Gas Law of the Kurdistan Region-Iraq, Law No. (22)-2007, available at www.krg.org/uploads/documents/Kurdistan%20Oil%20and%20Gas%20Law%20English_2007_09_06_h14m0s42.pdf (accessed Apr. 4, 2009).

Constitution that any treatment accorded by the KRG to its contracts executed since 1992 must be consistent with all other provisions of the Iraqi Constitution. The third observation is related to the precise reach of the concept of finality, as that concept is typically understood.[20]

Regarding the first observation – that is, the distinction concerning finality between the First and Second paragraphs of article 54 – it will be recalled that article 141 of the Iraqi Constitution, adopted in 2005, was preceded by various draft versions. Those drafts that were temporally most proximate to the penultimate version aimed to confirm the validity of preexisting KRG contracts, dependent on the KRG not desiring to regard them differently. Article 54 represents the KRG exercise of that constitutional authority by providing in the First paragraph, that agreements related to PSCs are essentially considered valid but subject to Regional Council review and possible alteration to ensure consistency with the many provisions of the KRG oil and gas law (No. 22). Decisions of the Regional Council consequent to such reviews – presumably even those resulting in alterations to the terms of earlier contracts – however, are deemed "final." The Second paragraph of article 54 indicates that the KRG chose to exercise its article 141 constitutional authority over preexisting authorizations and MOUs related to oil and gas in a distinctly divergent manner. Specifically, this entailed from the outset that all such preexisting contractual arrangements are regarded as null and void unless they are otherwise approved and validated by Regional Council action.

With respect to the article 54 invocation of finality, does it not make sense to characterize Regional Council decisions on PSC-related agreements as "final," yet to not so characterize those involving authorizations and MOUs related to oil and gas? For openers, regarding the latter, the authorizations and MOUs are statutorily designated as "null and void."[21] In other words, even though earlier negotiated and concluded with personnel appropriately exercising KRG governmental authority, and even though representing the sanctity of a solemn contractual arrangement, the legal instruments concerned are deemed

[20] The analysis in this section relies heavily on the textualist interpretive methodology, with reference to the total statutory context and provisions of article 54 of the Iraqi Constitution suggesting at least a nod in the direction of the teleological approach. See supra notes 9 and 11 and accompanying text.

[21] See art. 54, second, supra Chpt. 4, note 23.

wholly without any continuing consequence – invalid and at an end. Given the severity of this consequence – especially when contrasted with the significantly less ambitious reach of mere review and possible contractual modification accompanying preexisting PSC-related agreements – does it not seem reasonable to refrain from applying the concept of finality to reviews that refuse to reinvigorate dead arrangements, but not to those that simply seek to modify the terms of PSC-related agreements to ensure consistency with the provisions of the KRG oil and gas law (No. 22)? Is not the reasonableness of the conclusion associated with this query further supported by the fact that any modifications to the terms of agreements related to PSCs something to be advanced by the Regional Council experts? Surely, it would seem peculiar to subject the expert and politically attuned decisions to reconsideration by a legal body schooled only in the nuances and intricacies of the law.

The second principal observation about article 54 – regarding its interface with article 141 of the Iraqi Constitution – is concerned with the inference drawn from the statement in article 141 that any action taken by the KRG to treat contracts made since 1992 as other than of continuing validity, must be taken with an eye to respecting rights protected elsewhere in the Iraqi Constitution.[22] Without describing any of these rights, suffice it to state that if the reference in article 54, First, of the KRG's oil and gas law (No. 22) to Regional Council decisions on reviews of agreements related to PSCs being "final" meant to preclude any and all reexaminations of Regional Council decisions, then the protections incident to the constitutional rights mentioned in article 141 would be rendered hollow and meaningless. Rights specified elsewhere in the Constitution as protecting interests of holders of contractual arrangements in the form of PSC-related agreements would be entirely inconsequential; review decisions of the Regional Council that focus on such agreements would trump every constitutional protection and truncate any possibility of reconsideration. To avoid such an absurd result and to give real meaning to the article 141 reference to KRG actions about preexisting contracts being in accord with other provisions of the Iraqi Constitution – that is, to read articles 54 and 141 in a way that gives effect to both – the article 54, First, reference

[22] See supra Chpt. 4, note 3.

to "final" must be understood as at least permitting legal challenge based on Regional Council decisions not being in accord with relevant constitutional rights. Essentially, the effect is that the article 54, First, declaration of finality is not what it initially may appear.

The third observation concerns the common understanding of the concept of finality. To a certain extent this ties into the previous intimation that it seems strange to have Regional Council decisions about which emendations and alterations to make to preexisting PSC-related agreements be subjected to reexamination by a legal body anointed to engage in the exercise of reconsideration. The Regional Council – not some court of arbitral body – is the expert in the needs and requirements of KRG oil and gas arrangements. This surely suggests that the wisdom of any amendment or modification should not be subject to reexamination by any court or arbitral body. This is consistent with the way finality is usually and commonly understood. By the same token, however, that concept is not understood as precluding review or reexamination of questions about whether actions taken were within the scope of an acting entity's authority, the procedural regularity of the actions, or the possibility of such actions contravening constitutional rights and protections. Concerning the narrow question of whether (and if so, how) agreements related to PSCs should be amended or modified by the Regional Council, reexamination may not be available. In other words, actions of the Regional Council that involve the exercise of public-policy discretion and political judgment – as distinct from observance of legal constraints and dictates – cannot be subjected to reconsideration by any supervisorial entity. On a host of other matters, however, the opportunity to request reexamination remains intact. As long as any reexamination request is limited to these matters alone, then finality is fully respected, even though Regional Council decisions are scrutinized by an overseer.

III. THE MEANING OF "RELATED TO" AS USED IN ARTICLE 54

The language of article 54 of the KRG oil and gas law (No. 22) raises another interesting question in the use of the expression "related to."[23]

[23] See supra note 1.

Article 54, First, addresses Regional Council review of all agreements "related to" PSCs, and article 54, Second, addresses the null and void status of all authorizations and MOUs "related to" oil and gas. Furthermore, the definitional section of the oil and gas law indicates that "authorizations" mean petroleum contracts, prospecting authorizations, or any agreement "in respect of" such.[24] Clearly, the expression "related to" as used in both paragraphs of article 54 could have one of two distinct meanings, with the same being true of the expression "in respect of" appearing in the definition of the term *authorizations*. For openers, both expressions can be taken to mean that any and all agreements "in the nature of" or "in the form of" a PSC, a petroleum contract, an oil- and gas-prospecting authorization, and the like, or an MOU, are subject (where relevant) to Regional Council review for determining consistency with the terms and provisions of the oil and gas law, or to invalidation and nullification in accordance with the directive of that law. Comparatively speaking, such an interpretation of "related to" and "in respect of" advances a narrow and confined definition of either expression. A significantly broader and more encompassing interpretation exists, however: Those expressions assume that any and all arrangements "connected to" or "associated with" PSCs are subject to review, whereas arrangements "connected to" or "associated with" petroleum contracts, oil- and gas-prospecting authorizations, and the like, and MOUs are regarded as null and void. Regarding this latter interpretation, it is recognized that the words "related to" in article 54, Second's, reference to authorizations and MOUs "related to" oil and gas means those "connected to" or "associated with" the matter of oil and gas activity. The concern here, however, is whether the words also envision article 54, Second's, effect of nullity and invalidity operating on arrangements that are simply "in the nature of" or "in the form of" an oil and gas authorization or MOU?[25]

It is difficult to know with certainty which of the two suggested interpretations should be ascribed to the article 54 direct or indirect references to "related to" and "in respect of." However, there are

[24] See KRG oil and gas law (No. 22), id. at art. 1(22).

[25] The analysis in this section relies heavily on textualism (see supra note 9) but also on intentionalism (see supra note 10), at least to the extent that recourse to the October 2006 Draft Petroleum Act of the Kurdistan Region of Iraq is suggested to indicate the state of mind of the drafters of article 54.

various reasons to believe that viewing the expressions as limited to legal arrangements "in the nature of" or "in the form of" PSCs, petroleum contracts, oil- and gas-prospecting authorizations, and the like, and MOUs is better. In any event, it cannot be denied that whichever reading is argued for, the expressions can impact those entities that entered oil and gas arrangements with the KRG between 1992 and August 2007. Any arrangement that is not a classic PSC would raise questions regarding its treatment as merely subject to Regional Council review, or null and void for being an agreement or MOU related to oil and gas.

Twenty-five uses of the expressions "related to" and "in respect of" appear in the KRG oil and gas law (No. 22). For example, article 2, First, paragraph (b), declares that the provisions of the law apply to "all activities related to Petroleum Operations,"[26] with *petroleum operations* defined in article 1(17) as including oil and gas "prospecting, exploration, development, production, marketing, transport, refining," and so forth.[27] Similarly, article 2, Second, provides that in accordance with two provisions of the Iraqi Constitution, "no federal legislation, and no agreement, contract, memorandum of understanding or other federal instrument that relates to Petroleum Operations shall have affect except with the express agreement of the . . . Region."[28] Article 15, Fifth, references "in respect of" in language indicating that until a national Iraqi revenue scheme is implemented, KRG oil and gas revenues are to be held in two separate accounts: "one for Revenues from Petroleum Operations in respect of Current Fields (Current Fields Account); and one for Revenues from Petroleum Operations in respect of Future Fields (Future Fields Account)."[29] Admittedly, these and other uses of "related to" and "in respect of" suggest that the expressions should be interpreted throughout the various provisions of the law – including in article 54 and its affiliated definitional provision – as the broader "connected to" or "associated with," rather than the narrower "in the nature of" or "in the form of." Nevertheless, there are two significant problems that prevent a broader interpretation to prove compelling in

[26] See id. at art. 2, First,(b).
[27] See id. at art. 1(17).
[28] See id. at art. 2, Second.
[29] See id. at art. 15, Fifth.

connection with article 54 and its necessary reference to the definitional provisions of the law.

The first problem concerns the context in which "related to" and "in respect of" are used either directly or indirectly in article 54. In particular, the way the KRG oil and gas law is written indicates that when the article references PSCs, petroleum contracts, oil- and gas-prospecting authorizations, and the like, or MOUs, it is limited to only those. Similarly, when it references legal arrangements "related to" or "in respect of," it means only arrangements that are in the "nature of" or the "form of" a PSC, petroleum contract, oil- and gas-prospecting authorizations, and the like, or MOUs – not those that may be somehow "connected to" or "associated with" them. Chapter Ten of the KRG oil and gas law (No. 22), entitled "Production Sharing Contract," provides the Ministry of Natural Resources with authority concerning PSCs, petroleum contracts, and oil- and gas-prospecting authorizations. That same chapter, however, also provides the Ministry with authority over so-called Access Authorizations[30] and, under article 39 of the law, a host of other forms of agreement used in the oil and gas sector – from service contracts to field management contracts, supply and installation contracts, construction contracts, consulting contracts, and others.[31] From the precise terms of article 54, however, the review and nullification authorities of the article are confined to particular identified forms of legal arrangement. Any form of arrangement not explicitly mentioned, therefore, falls outside the article's reach. Thus, by article 54 directly referencing agreements related to PSCs, authorizations and MOUs related to oil and gas, and – through the law's definition of "authorization[s]" as meaning petroleum contracts and oil- and gas-prospecting authorizations and the like – it does not include any other form of oil- and gas-development arrangements.

Separate from the interpretive significance of the overall structure of the KRG oil and gas law (No. 22), a second problem with a broad reading of "related to" or "in respect of" is the evolutionary background of the proposals that resulted in the language of article 54. The immediate draft predecessor version of the language that ultimately became article 54 appeared in the October 2006 Draft Petroleum Act

[30] See id. at art. 25.
[31] See id. at art. 39.

of the Kurdistan Region of Iraq.[32] Article 79 of that draft explicitly affirmed the continuing forcefulness and validity of every oil and gas agreement previously entered into by the KRG.[33] Section 1 of article 79 states: "Any agreement related to Petroleum Operations entered into by the Regional Government prior to the entry into force of this Act, and approved by the Minister, shall remain in force."[34] Section 2 of the same article then defined "agreement" to "include[] a contract, license, permit, memorandum of understanding, or *other legal act or dealing of any sort*."[35]

It is clear that this language was all-encompassing; it covered not only specifically named forms of legal arrangement (i.e., contracts, licenses, permits, and MOUs) but also all "other legal act[s] or dealing[s] of any sort." In opting to reject the route proposed in the language of draft article 79, article 54 of the August 2007 KRG oil and gas law (No. 22) signaled that the KRG preferred to move away from confirming the validity of all preexisting legal arrangements affecting oil and gas approved by the Minister of Natural Resources. However, instead of moving to a position of subjecting all such arrangements to review or nullification, the KRG simply chose to review or nullify specifically identified forms of oil- and gas-development arrangements. Presumably, the drafters knew how to signify that all such arrangements were reviewable or nullified, since in the earlier draft it is obvious that they knew how to provide that all agreements approved by the Minister of Natural Resources remained valid. Having chosen another formulation in the 2007 legislation, why should not that formulation be honored by those called on to apply the terms of the oil and gas law?[36]

32 See Draft Petroleum Act of the Kurdistan Region of Iraq, 22 Oct. 2006, available at www.krg.org/pdf/Kurdistan_Act_COM_draft_22_October_2006.pdf (accessed Apr. 4, 2009).
33 For the text, see id. at art. 79.
34 See id. at Sec. 1.
35 See id. at Sec. 2 (emphasis added).
36 It is to be acknowledged that the text of the proposed article 79, in referencing "approved by the Minister," can be read as referring to either those agreements approved by the Minister when entered into or those already entered into that the Minister approved after the passage of the KRG oil and gas law. In either case, it is clear that the definition of *agreement* as contained in the proposed article 79, Second, is all-inclusive, encompassing every conceivable form of oil and gas agreement, whereas the language of article 54 of the 2007 KRG oil and gas law references only specific types of agreements. One should also consult the text of Section V ("Conclusion") of this chapter, and its discussion of the fact article 54's use of "related to" might be seen (though not persuasively) as meaning "associated with" or "connected to," since the issue of third-party subcontracts

Why should the language of article 54 be construed to mean anything other than the PSCs, petroleum contracts, oil- and gas-prospecting authorizations, and the like, and MOUs, or other arrangements "in the nature of" or "in the form of" such are rendered reviewable or null and void?

IV. SUBCONTRACTS AND OTHER THIRD-PARTY ARRANGEMENTS

It is one thing to struggle with the concepts of "final[ity]" and "related to" in article 54, and something entirely different to attempt to divine whether the specifics of that article apply only in regard to oil and gas legal arrangements between the KRG and parties with which it has directly contracted, or also to secondary or tertiary arrangements that the contractors may enter into with various third-party servicers or associates. After all, there is concern about the question of the applicability of article 54 to technical service agreements (TSAs) or engineering, procurement, and construction (Eng.&PC) contracts between the KRG and a hypothetical international oil company (IOC), given the article's reference to agreements "related to" PSCs, petroleum contracts, oil- and gas-prospecting authorizations, and the like, and MOUs. However, nothing in the language of that provision or its drafting history proves helpful in determining the applicability to subsidiary arrangements between an IOC that contracted with the KRG and a third party that agrees to provide a service necessary to fully perform the oil company's contract.

To illustrate, assume that Geysera, Inc., entered into a PSC with the KRG in 2006 to develop a specifically identified oil field. Geysera subsequently contracts with Geotechco to provide geophysical data and analysis concerning the oil field to which the PSC relates. In the months after adoption of the oil and gas law (No. 22), the KRG Regional Council reviews Geysera's PSC and secures modifications to bring it into conformity with the terms and provisions of the new law. The modifications, however, result in consequences that make it

(examined in Section IV of this chapter) suggests those contracts should be subject in an indirect and trickle-down way to the application of the article's modification or nullification provisions.

impossible to fulfill the exact terms of the Geysera/Geotechco contract. Geotechco then institutes an action against Geysera contending breach of contract and demanding compensation. Geysera maintains that just as it would have been confronted with the article 54 reference to finality had it directly challenged the PSC modifications advanced by the KRG subsequent to Regional Council review, Geotechco encounters the same impediment in its action involving the subsidiary service contract. Geotechco replies that its service contract could not have been directly affected by the terms of article 54 because that contract was neither a PSC, a petroleum contract, an oil- and gas-prospecting authorization, nor the like, nor a MOU; therefore, it cannot be indirectly affected through connection with the Geysera PSC. Geysera responds that Geotechco's contract is an integral appendage and outgrowth of the PSC with the KRG and, thus, is as equally subject to the aspects of article 54 as the PSC itself.

This hypothetical scenario raises the matter of the extent of the article 54 reach. Is it limited to those legal arrangements between the KRG and direct contracting parties? Or is it capable of extending to arrangements with third parties as well? While the context of article 54, its drafting history, its relationship to article 141 of the Iraqi Constitution, and its setting in the overall structure of various other provisions of the KRG oil and gas law (No. 22) may provide assistance in understanding the concepts of "final[ity]" and "related to," what exists to offer insight about the article 54 affect on subcontracts and other third-party arrangements? Should it simply be assumed that just as the direct and immediate legal arrangements between the KRG and an IOC fall within the ambit of the article 54 review and nullification authorities, so too does any subsidiary or tertiary arrangement of a third-party servicer or associate that is dependent on the existence of the KRG/IOC primary contract? Or, conversely, should any such subsidiary or tertiary arrangement be viewed as wholly standing on its own and left entirely beyond what article 54 is capable of affecting?

Starting from the perspective of practicality, there can be no doubt about a paucity of language in article 54 offering insight to the matter of its reach to third-party legal arrangements; however, it would seem somewhat unreasonable to view that as not within the article's capacity. Imagine the ramifications to third-party agreements, if the terms of a principal legal arrangement between the KRG and an IOC could be

reviewed by the Regional Council and subjected to substantive modifications to ensure conformity with the provisions of oil and gas law (No. 22) or declared null and void because it is a mere article 54, Second, authorization or MOU related to oil and gas. How sensible is it to insist that all third-party arrangements predicated on a principal agreement deemed null and void remain in place? By virtue of the elimination of the *raison d'être* for the subsidiary or tertiary arrangements, should they not also face elimination? In the absence of the continuation of the KRG/IOC commitment, the need or rationale for the continuation of mere supportive arrangements would vanish. In the event that the principal KRG/IOC commitment were altered simply to guarantee consistency with the terms of the KRG oil and gas law (No. 22), would it make sense to view potentially inconsistent provisions of third-party arrangements as frozen and unchanging? Any interpretation of article 54 producing such a result would generate a situation in which those obligated under the subsidiary or tertiary arrangements are confronted with choosing between acting in contravention of KRG law and complying with the specific terms of their contractual arrangement.

Regarding any inclination toward insisting that, no matter what reason may suggest, a contract is a contract and, therefore, what is required by the terms of a subcontract must be fulfilled to the letter despite what may happen to any principal contract with which a subcontract is connected, a modicum of support for the contrary position can be grudgingly extracted from the otherwise virtually barren language of article 54. Those who venerate the sanctity of contractual commitment and are equally as insistent that any Regional Council alteration in a principal contract governing the relationship between the KRG and an IOC does not trickle down and correspondingly affect subcontracts, cannot avoid the reality of the article 54 distinction in the treatment between agreements in the form of PSCs and those in the form of petroleum contracts, oil- and gas-prospecting authorizations, and the like, or MOUs. Pursuit of the preservation of contract above all else meets with stark reality in that article's clarity that agreements in the form of PSCs are subject to no more than modification in the context of Regional Council review, whereas those in the form of petroleum contracts, oil- and gas-prospecting authorizations, and the like, or MOUs are seen as null and void. To devotees of the sanctity of contract, a

kernel of appeal may be evidenced in the argument that modifications to a principal contract should in no way affect associated subcontracts. However, even those proudly bearing that stripe must acknowledge that the fact article 54 distinguishes between the treatment accorded agreements in the form of PSCs and those in the form of petroleum contracts, oil- and gas-prospecting authorizations, and the like, and MOUs confronts one with a substantial discontinuity. Specifically, while it may not seem entirely untenable to contend that the terms of a subcontract should be respected as written – even though an associated principal contract is modified as a result of Regional Council review – when considering a subcontract pegged to a principal legal arrangement deemed by article 54, Second, to be null and void, that same contention loses much of its cachet. Accepting the view that a subcontract remains, even when the principal arrangement with which it is associated has disappeared, leaves the unsatisfactory situation of a subsidiary or tertiary servicer or associate with the ability to insist on the performance of its contract when the KRG/IOC legal deal on which it hinges has vanished. Although article 54 may contain nothing that explicitly addresses its reach to third-party subcontracts, its distinction in treatment among various types of principal contract arrangements suggests that it envisions extension to subcontracts as well.

There is also another reason for construing article 54 as not being limited in its reach to just principal contract arrangements between the KRG and an IOC: construing that article to have the opposite effect serves to incentivize IOCs to subcontract out all important duties. In doing so, the IOC could claim to avoid attempted modifications or efforts to designate particular agreements as null and void. With article 54 reaching no farther than the original master contract (so to speak), desired modifications or declarations of invalidity would prove inconsequential for contractual arrangements once removed. From the vantage of protecting public policy, this would seem counterproductive. Although there may be provisions of the Iraqi Constitution that protect property rights inherent in contractual commitments (see Chapter 6), thereby otherwise acting to circumscribe or limit the authority of the KRG in its exercise of power under article 54 of the oil and gas law (No. 22), it is unpalatable to interpret the language of that article in a way that encourages subcontracting as a stratagem for short-circuiting expressions of public policy by legislative and quasi-legislative bodies.

By subjecting both the principal contract and all subsidiary or tertiary arrangements to the terms of article 54, the ability of the KRG to give full voice to its public-policy choices is preserved.

Finally, two other matters merit consideration with regard to whether changes in the original state of affairs concerning the so-called master legal arrangement between the KRG and an IOC trickle down and affect any subcontracts or other third-party arrangements: (1) the notion that the very terms and provisions of any arrangement between an IOC and a service provider or associated company might speak to that matter; and (2) the generally recognized legal defenses of impossibility or impracticability. With respect to the former, the idea would be not that the terms and provisions of any subcontract or third-party arrangement purport to claim some tenuous contractual basis bypassing the effect of article 54 – for that would not prove doable – but rather that the language of the terms and provisions of any arrangement specify the precise consequences to flow from an article 54 review and modification or a nullification and invalidation. If the affiliated subcontracts or third-party arrangements specified what was to proceed from such an event, then there would be little concern about the trickle down potential of changes in a KRG/IOC contractual arrangement. And with respect to the defenses of impossibility or impracticability, these may be capable of invocation by the IOC to avoid arguments of subcontractors or third parties that the IOC can escape the terms of its subsidiary legal arrangements only at the risk of damages for breach of contract. Access to those defenses – which emerge from the fact that the main contractual undertaking on which the subcontracts or third-party arrangements rest has been changed or affected by sovereign legal mandate – might secure for the IOC safe harbor from any effective breach of contract claim. Although the complexities related to impossibility or impracticability are widely acknowledged,[37] both have found traction in domestic and international law circles.[38]

[37] See, e.g., J. White & R. Summer, Uniform Commercial Code 143 (5th ed., 2000).

[38] See, e.g., American Law Institute, Restatement (Second) of the Law of Contracts, Sec. 264 (1981); Uniform Commercial Code, 2–614, available at www.law.cornell.edu/ucc/2/overview.html (accessed Apr. 18, 2009); United Nations Convention on Contracts for International Sales of Goods, art. 79 ("impediment"), available at www.cisg.law.pace.edu/cisg/text/e-text-79.html (accessed Apr. 18, 2009); Robert Y. Jennings, "Chpt. 6: Treaties," in *International Law: Achievements and Prospects* 159–60 (ed. Mohammed Bedjaoui, 1991).

If alterations or consequences are imposed on the master contract by clear and unequivocal legal directive, then it is peculiar to think that the terms of any dependent and associated legal arrangement based on the master contract could be insisted on even though doing so would lead to actions that contravene the KRG law.

V. CONCLUSION

Some comments are in order about two matters before Chapter 6 discusses the provisions of the Iraqi Constitution that restrict or limit the KRG's ability to exercise article 54 powers in a way that proves too abusive of preexisting contractual rights, since it is obvious that the various intricacies of article 54 could surface in the context of a legal claim made against Iraqi oil and gas or revenues from their sale. These two matters concern: (1) the relationship between the observations just advanced in connection with third-party subcontracts and the earlier observations about article 54's use of "related to" as meaning "in the nature of" or "in the form of"; and (2) the matter of so-called stabilizations provisions.

Regarding the relationship between third-party subcontracts and article 54's use of the words "related to," it will be recalled that the narrower meaning of those words as "in the nature of" or "in the form of" a PSC, oil and gas authorization or MOU, rather than as "associated with" or "connected to" such, argued against the KRG being able to review and modify, or consider as null and void, its agreements that took the form of something other than such (e.g., TSAs, Eng.&PCs). As just seen in the immediately preceding section on third-party subcontracts, however, in the event the KRG has a principal or master contract that is a PSC, oil and gas authorization or MOU, then any associated or affiliated third-party arrangements had between the IOC and subcontractors would seem subject to changes or invalidations in line with changes or invalidations made in the principal or master KRG/IOC agreement.

It cannot be stressed strongly enough that all such changes or invalidations in such third-party subcontracts are to eventuate not from action by the KRG directly against those associated or affiliated with the IOC through third-party subcontracts, but rather through action

that touches such subcontractors, if at all, only indirectly after trickling down from the KRG/IOC master contract being modified or nullified in a way inconsistent with the continuation of the exact same terms of any subsidiary contracts. The end result might be the same as would be reached through according article 54's words "related to" the broader reading of "associated with" or "connected to." It would be reached, though, by virtue of the IOC raising the matter of contract change or nullification, not the KRG directly confronting third-party subcontractors with modifications or the claim of contract nullification. While allowing for this sort of distinction between indirect as opposed to direct modification or nullification could be seen as undercutting the basic notion advanced in Section III that article 54's words "related to" are best not construed as meaning "associated with" or "connected to," a sound reason for the distinction may reside in the fact that, when article 54's two paragraphs are read in light of its predecessor, article 79 of the KRG's 2006 Draft Petroleum Act, it has an important effect. Specifically, the provisions permit KRG Regional Council alterations to PSCs and agreements in the form of such, to nullify and invalidate authorizations and MOUs connected to oil and gas activities, but leave in place and confirm the legitimacy of any and all other forms of oil and gas activity arrangements earlier struck by the Kurds directly with other contracting partners. Oil and gas activity arrangements between third-party subcontractors and those with whom the Kurds have struck something like master PSCs or MOUs, for example, would remain subject to the legal vagaries growing out of the trickle down effect of article 54's modification and nullification provisions.

Now with respect to the matter of "stabilization provisions," these are provisions contained in international contracts between host-country governments and private contracting parties, and they aim to circumscribe the ability of the host government to exercise its legislative or rule-making power in a way that changes the legal environment subsequent to the terms of a contract being "cast in stone."[39] Clearly, were a stabilization provision to appear in a principal contractual arrangement between the KRG and an IOC, the idea is that the article 54 Regional Council review and modification authority – that is,

[39] See, generally, A.F.M. Maniruzzaman, "The Pursuit of Stability in International Energy Investment Contracts: A Critical Appraisal of the Emerging Trends," 1 *J. World Energy Law & Bus.* 121 (2008).

its designation of specific kinds of IOC arrangements as null and void – would not have any consequence for subcontracts or third-party arrangements. Because the stabilization provision preserves the integrity of the terms of the original KRG/IOC commitment, all terms of subsidiary or tertiary arrangements with subcontractors or associated entities would also be left in place.

To be sure, the particular effect of the stabilization provision on the principal contract – and, therefore, all related contracts – hinges on the precise language of that provision. If, for instance, the language were phrased such that it only served to stabilize or freeze the effect of KRG law on specialized and particular aspects of the KRG/IOC master commitment not relevant to rights and duties addressed in the subsidiary or tertiary arrangements, then the existence of the stabilization provision would prove of no moment for the continuing validity of the third-party arrangements as written. Modifications made to or the nullification and voidance of master-contract terms would pass through to apposite subcontractors consistent with the analysis in Section IV. Conversely, if the stabilization provision used language that either (1) spoke explicitly to rights and duties in the master commitment that were replicated in the legal agreements between third parties and the IOC, or (2) was voiced in such a general and encompassing way as to reveal the intent that every aspect of the KRG/IOC contract would remain untouched by changes of law subsequent to contract finalization, then the stabilization provision would be meaningful in protecting subcontractor rights and duties as evidenced in the terms and provisions of subsidiary or tertiary arrangements with the IOC. After all, if a stabilization provision insulates some or every aspect of a master contract with the KRG from changes in the law made subsequent to the contract being entered into, would it not seem reasonable to view pertinent, connected aspects of third-party arrangements grounded on that master contract as similarly unaffected by those changes in the law?

As mentioned in the preceding section, the precise language of the terms and provisions of any subsidiary or tertiary arrangement is not without significant relevancy. In regard to the idea of stabilization provisions, it is always possible that they may appear in a principal or master commitment between the KRG and an IOC. At the same time, the language of any IOC/third-party arrangement may provide expressly that KRG law enacted subsequent to the entering into of that

arrangement – including a law such as article 54 of the August 2007 oil and gas law (No. 22) – operate to affect the nature of the rights and duties under that arrangement. Without any firsthand knowledge about whether any of the preexisting master contracts with the KRG, or subcontracts with relevant IOCs, contain either stabilization provisions or provisions addressing the matter of whether subsequent legal enactments impact master or subcontract rights and duties, it would not seem at all unusual for the KRG to want to keep some control over subcontracts associated with its principal or master contracts. To accomplish this objective, the KRG occasionally may have leaned on IOCs to insert in subcontracts some such language. Whether this ever happened is speculation. Nonetheless, it would seem pretty clear that if it had, any language in a subsidiary or tertiary arrangement linked to the performance of an IOC master contract, and declaring that the effect of a stabilization provision in the master does not protect that arrangement from later legal enactments, would be deliberately designed to assert KRG sovereign authority to claim all commitments between private and other parties are subjected to current (and not earlier) legal directives. This may appear inconsistent with the general notion that what is in the master contract percolates to all subsidiary- or tertiary-related agreements. However, it is the fact that those agreements contain specific consented-to contractual language to the contrary that explains this as an inconsistency in appearance alone, not in reality.

6 OLD OIL AND GAS CONTRACTS AND THOSE OF MORE RECENT VINTAGE: IRAQI CONSTITUTIONAL PROTECTIONS OF CONTRACT RIGHTS AND THEIR SIGNIFICANCE ON LEGAL CLAIMS

I. INTRODUCTION

As discussed in Chapter 5, the full import of KRG oil and gas law (No. 22) article 54 cannot be accurately gauged without considering the limitations imposed on that provision by the overarching and superior provisions of the Iraqi Constitution. It is one thing to speculate about the significance of the article 54 references to "final[ity]" and "related to" or "in respect of," and about the chances that the article's reach encompasses third-party contractual arrangements that assist the IOC in meeting its obligations with the KRG. It is entirely different, however, to begin to "wrap one's mind around" the exact scope of any KRG assertion of power under article 54 in light of the various limitations found in the terms of the Iraqi Constitution on all exercises of governmental power, whether at the central or subcentral level. The easiest way to conceptualize this entails recalling the article 54 claim to vest the Regional Council with the authority to review and alter all preexisting agreements related to PSCs entered into by the KRG prior to August 2007, and its pronouncement that authorizations and MOUs related to oil and gas entered in that period are deemed null and void. The obvious thrust of that article is to affect preexisting oil and gas legal arrangements in one way or another. Whether that objective is capable of unfettered effectuation, however, hinges on the

presence of constitutional protections designed to preserve rights incident to contractual obligations.

Before referring specifically to such constitutional protections, there is a second matter that bears noting, given that the two together form the basis of this chapter. Specifically, the second matter concerns the fact that the contrast between the language of article 54 of the oil and gas law (No. 22) and article 141 of the Iraqi Constitution suggests oil and gas contracts could fall into two distinct categories. Here, reference is not made to article 54's explicitly stated categories of, on the one hand, PSC-related agreements and, on the other, authorizations and MOUs related to oil and gas. Rather, the reference is to oil and gas contractual arrangements entered into by Kurdish regional authorities before 1992 and those concluded thereafter – whether between the time when the Kurdistan Region was accorded the protection of the coalition's northern no-fly zone and Saddam Hussein's removal from power in spring 2003, or between the time of his removal and the adoption of the August 2007 KRG oil and gas law (No. 22).

Unlike article 54 of the oil and gas law (No. 22), which is written only in terms of agreements entered into "by the Regional Government prior to the entry into force of this Law," article 141 of the Iraqi Constitution makes a clear distinction between what might be referred to as the older 1992 to spring 2003 oil and gas contractual arrangements, and those entered into during the more recent period from spring 2003 to August 2007. Article 141 confirms the continued validity of contracts concluded by the KRG since 1992, unless altered by actions of the KRG after adoption of the Constitution. Thus, the language of article 141 distinguishes pre-1992 contracts from those entered into after 1992. In contradistinction, article 54 of the oil and gas law (No. 22) is written to suggest that no difference in treatment is to be accorded oil and gas legal arrangements simply because they date from a period beyond 1992, rather than from post-1992 to the removal of Saddam Hussein in spring 2003, or from the period after his ouster until August 2007.

This discussion illustrates that there is no possibility of avoiding the conclusion that a reading of article 141 of the Iraqi Constitution, in conjunction with article 54 of the KRG oil and gas law (No. 22), presents the question of limitations that exist in the Constitution to restrict the KRG's exercise of authority to determine how preexisting

oil and gas contracts should be treated, as well as that of whether any distinction exists between contracts reportedly concluded prior to 1992, and those concluded thereafter. Article 141 of the Constitution is explicit in dictating that how the KRG exercises the authority granted by that article must be in accord with all other terms of the Iraqi Constitution. At the same time, it contrasts with article 54 of the KRG oil and gas law (No. 22) in referencing nothing other than the KRG's ability to regard post-1992 contracts as subject to KRG declarations of validity or invalidity. Article 54 may contain language that can be interpreted as applicable to pre-1992 contracts, but no comparable language appears in article 141 of the Iraqi Constitution.[1]

II. OIL AND GAS CONTRACTUAL ARRANGEMENTS OF THE KRG CONCLUDED BEFORE OR AFTER 1992

The difference in the language of article 54 of the KRG oil and gas law (No. 22) and that of article 141 of the Iraqi Constitution raises the question about whether any oil and gas contractual arrangements reportedly negotiated by the KRG before 1992 remain valid. (It is unlikely that any commitments exist, given the fact that until 1992 and the establishment of the northern no-fly zone, no formal, independent, and recognized governing structure in the Kurdistan Region is known to have continuously and effectively operated.) Article 141 of the Constitution was adopted in 2005, several months prior to even the October 2006 Draft Petroleum Act of the Kurdistan Region of Iraq. It is therefore understandable that the language of article 79 in the proposed draft[2] suggested the continued validity of all ministerial-approved pre-existing oil and gas agreements, whether concluded before or after 1992. The precise terms of article 141 of the Constitution state that legislation adopted in the Kurdistan Region since 1992 and decisions of the KRG, including court decisions and "*contracts*, shall be considered valid unless they are amended or annulled pursuant to the laws of

1 At this juncture, recall that as the discussion in Chpt. 5, supra, indicated, various interpretive methodologies exist in the context of attempting to decipher the meaning of confusing or ambiguous legal provisions. See supra Chpt. 5, notes 9–13 and accompanying text. Comment offered previously about these methodologies also has relevance in the present chapter.

2 See supra Chpt. 5, text accompanying notes 32–35.

the region . . . by the competent entity in the region, provided that they do not contradict with the Constitution."[3] It is clear that draft article 79 of the proposed October 2006 Draft Petroleum Act was designed to follow this at the regional level with language providing that any KRG agreement related to petroleum operations – whether labeled a contract, license, permit, MOU, or anything else – "shall remain in force" if approved by the KRG Minister of Natural Resources; no linkage was required by draft article 79 to a particular date.[4]

Adoption of the particular language of article 54 of the 2007 KRG oil and gas law (No. 22) evidenced a clear retreat from the exercise of the full power provided the KRG by the Constitution, as well as from the approach of draft article 79 in the 2006 proposal. The Constitution had authorized the validation of all KRG contracts negotiated since 1992, and article 79 of the proposed October 2006 Draft Petroleum Act of the Kurdistan Region of Iraq seemed to continue that line as well. Article 54 of the KRG oil and gas law (No. 22), however, raised the prospect of validating only some forms of contract and invalidating many others. Nevertheless, by virtue of the Constitution's supremacy over regional law, the KRG choice to review and change some preexisting legal arrangements seen as of continuing validity, and totally invalidate and void others, remained subject to serious limitation. The significance of this limitation – expressed in the straightforward language of article 141 and calling on other constitutional provisions protecting preexisting rights – forms the basis of much of the discussion in subsequent sections. However, the concern in this section is whether any distinction should be made between oil and gas contractual arrangements entered into by the KRG before and after 1992. Article 54 suggests a negative answer, whereas the language of article 141 of the Constitution suggests an affirmative answer.

Examining that matter more closely, it is indisputable that article 54 does not link its approach to a particular date on which oil and gas legal arrangements were entered into, whereas article 141 does. Other than the previous mention to the fact that it is unlikely that any pre-1992 KRG oil and gas legal arrangements exist, there are a couple of significant reasons for interpreting article 54 of the oil and gas law

[3] See supra Chpt. 4, text accompanying note 3 (emphasis added).
[4] See draft art. 79 of 2006 Petroleum Act of the Kurdistan Region of Iraq at supra Chpt. 4, note 28.

(No. 22) to have exactly the same linkage to 1992 as article 141 of the Iraqi Constitution. This would frustrate suggestions of contract validity if pre-1992 arrangements were ever to surface. The first significant reason for reading article 54 as linked to 1992, although it fails to so state, is related to the supremacy of the Iraqi Constitution over regional law when it comes to KRG contracts referenced in article 141. Article 141 provides that "[l]egislation enacted in the region of Kurdistan since 1992 shall remain in force, and decisions issued by the government of the region of Kurdistan, including . . . contracts, shall be considered valid . . ." unless amended or annulled by the KRG, consistent with the Constitution. By these terms, article 141 apparently ensures the continuing validity of legislation, court decisions, and governmental contracts adopted, issued, or negotiated by the KRG since 1992. Article 141 vests the KRG with authority to disavow such legislation, court decisions, or governmental contracts as long as the disavowal is consistent with other constitutional provisions. The language of article 54, Second, of the KRG oil and gas law (No. 22) that deems authorizations and MOUs related to oil and gas null and void unless otherwise approved by the Regional Council is an exercise of that constitutional authority. The language in article 141, however, does not entitle the KRG to declare pre-1992 contractual arrangements to be valid; yet, that is precisely what article 54 of the KRG oil and gas law seems to attempt.

Continuing with the idea of the supremacy of article 141 of the Constitution over article 54 of the KRG law, it cannot be denied that the exact language of article 141 raises interesting questions regarding intent. Specifically, it clearly references 1992 in regard to "[l]egislation," but in the follow-on reference to "decisions" by the KRG, "including court decisions and contracts," there is no such explicit reference. Thus presented is the question of whether article 141 links legislation only to 1992 and purports to leave in force all KRG court decisions and contracts, no matter when they were earlier issued or negotiated. In addition, that same language refers only to KRG "contracts," stating nothing about legal arrangements that might have been styled as something other than a contract (e.g., an authorization or MOU). Is it the intent of such language to apply article 141's validation of continuing force only to "contracts," and not other forms of contractual commitments?

On both of these matters, the negotiating background of article 141 (see Chapter 4, Section II) is extremely relevant. Concerning the explicit linkage of 1992 to KRG "legislation," it must be noted that the history of article 141's development can be interpreted as suggesting that no particular significance should be drawn from the explicit linkage of 1992 to legislation alone. The two most immediate predecessors of article 141 (i.e., article 137 of the August 28, 2005, draft Constitution[5] and article 150 of the August 23 draft[6]) – both seemed to tie the year 1992 to legislation alone. However, the seminal draft version of what was ultimately to evolve into article 141 – that is, article 80 of the July 28, 2005, Kurdish draft of the Iraqi Constitution – stated simply that "executive, legislative, and judicial authorities of the Iraqi Kurdistan Region are considered as legitimate as from the time of their establishment in the year of 1992."[7] Although the language designed to capture the sentiment of the July 28 draft was reformulated in the process, nothing suggests that it consciously aimed to link 1992 with KRG legislation alone, which leaves the potential for the argument that KRG actions of a judicial or executive nature (i.e., court decisions or contracts) predating 1992 might be viewed as of continuing validity. As for the article 141 use of the term *contracts*, when it comes to validation of continuing force of certain preexisting KRG legal arrangements, that same evolutionary negotiating background is relevant. It is clear from the language of article 80 of the Kurds' July 28 draft of what ultimately became article 141 of the Iraqi Constitution that the idea was to validate the continuing force of all executive actions taken since 1992 by KRG authorities, without regard to whether a contractual action was labeled a contract or labeled something else. Further evidencing this is the fact that even the subsequent draft article 150 of August 23 spoke of validating all *agreements*, a term more than adequate to extend well beyond contracts (strictly speaking). Moreover, there is ample reason to believe that the very term *contracts*, as it appears in article 141, is sufficiently broad to include all forms of contractual

[5] See supra Chpt. 4, note 16 ("Laws legislated in Kurdistan since 1992 remain in effect, and decisions made by the government of the Kurdistan region – including contracts and court decisions – are effective. . . .").

[6] See id. at note 15 ("Legislation enacted in the region of Kurdistan since 1992 shall remain in force. Decisions issued by the government of the region of Kurdistan, including court decisions and agreements, shall be considered valid. . . .").

[7] See id. at note 13.

undertakings, no matter whether they are styled as a contract, an agreement, an authorization, a license or permit, an MOU, or anything else.

Moving away from the supremacy of article 141 of the Iraqi Constitution, the second reason for interpreting article 54 of the KRG oil and gas law (No. 22) as tied to 1992 – although it does not contain any language making that connection clear – is concerned with the fact that article 54 really constitutes more of an expression about how the KRG has chosen to exercise its article 141 constitutional grant of authority to validate (or not) the continuing force of earlier contracts, rather than a signification of conflict with article 141 over the forcefulness of pre-1992 contracts, if any. Article 54 is cast in language that circumscribes its reach to agreements related to PSCs, and to authorizations and MOUs related to oil and gas. Article 141 of the Iraqi Constitution, by contrast, is not confined to contracts, but rather extends also to KRG legislation and court decisions as well – in other words, legislative-, executive-, and judicial-branch exercises of authority. Article 141 empowers the KRG to regard all such exercises of authority since 1992 as of continuing validity, or to amend or annul them. With article 54 of the oil and gas law (No. 22), the KRG can reasonably be assumed to have done nothing more than reveal a desire to distinguish between the treatment it intended to accord certain forms of contractual arrangements. Understandably, preexisting legislation and court decisions might not be the subject of an oil and gas law, but it is conceivable that the law's distinction between some contracts considered valid but reviewable, and others deemed invalid, implies that the real thrust of the article 54 language has nothing at all to do with how pre-1992 contractual arrangements are to be treated. Accepting this, it is inappropriate to infer that a side-by-side reading of articles 54 and 141 leads to the conclusion that the latter is confined to acknowledging the potential validity of KRG contracts entered into since 1992, whereas the former extends also to any existing pre-1992 contracts.

Regarding this second reason for maintaining that article 54 of the KRG oil and gas law (No. 22) should not be interpreted as validating pre-1992 Kurdish contractual arrangements – despite the fact that a side-by-side reading of article 54 and article 141 could lead to the opposite conclusion – another important but only related point bears mention. That point has to do with the negotiating background of article 54 and the previous observation regarding the move in the article 141 draft versions; a move from clearly tying all Kurdistan Region exercises

of legislative, executive, and judicial authorities to 1992, to an apparent distinction between the importance of that date and "legislation," on the one hand, and "court decisions and contracts," on the other. Should not the fact of the interpretive significance of evolutionary changes in the language of what became article 54 of the KRG oil and gas law (No. 22) carry over and affect how evolutionary changes in what became the final language of article 141 of the Iraqi Constitution are viewed? Is there not a certain inconsistency in arguing that changes in the drafts that led to article 141 prove inconsequential, yet changes in what led to the penultimate version of article 54 prove of great consequence?

As discussed previously, the move from what the Kurds had originally proposed in article 80 of their July 28, 2005, draft of the Iraqi Constitution, to the language of the completed article 141, was not viewed as significant, even though it involved a move to plainly different language.[8] One reason that the language shift was played down concerned the original intent behind what became article 141, which does not seem to have changed since its seminal predecessor was proposed. In regard to article 54 of the KRG oil and gas law (No. 22), its real thrust can be seen as having nothing to do with representing a challenge to article 141 of the Constitution over the matter of pre-1992 contractual arrangements. Rather, it reveals how the KRG chose to exercise constitutional discretion regarding which types of arrangements to validate and which to nullify; the language used to reflect the thrust of article 54 also moved from what had been originally proposed. As discussed in Chapter 5, the predecessor to article 54 was article 79 of the October 2006 Draft Petroleum Act of the Kurdistan Region of Iraq, and it provided for the continuing validity of all oil and gas arrangements, regardless of form. In particular, section 1 of article 79 indicated that "[a]ny agreement related to Petroleum Operations" remained valid; section 2 defined "agreement" to mean "a contract, license, permit, memorandum of understanding, or other legal act or dealing *of any sort*."[9] Chapter 5 argues that this move from the draft article 79 formulation to that settled on in the text of article 54 was significant, resulting in the article 54 references to "agreements related to Production Sharing Contracts" and "authorizations and memoranda

[8] See supra Chpt. 4, Sec. II, especially notes 12–13.
[9] See supra Chpt. 5, notes 34–35 (emphasis added).

of understanding related to oil and gas" being circumscribed in the types of legal arrangements reached.[10] By this measure, the meaningfulness of the article 54 shift in language suggests a similar result regarding the article 141 shift.

What operates to prevent that effect in this case, however, is that the shift to the language of article 54 from that of the article 79 predecessor seems intended to move from validation of all forms of KRG legal agreements to only those of a certain type. However, the shift involving article 141 of the Iraqi Constitution seems not as plainly intended to link KRG legislation – as distinct from court decisions and contracts – to the 1992 date. It is one thing to explicitly state that "any agreement" of "any sort" is valid, and then later move to language that distinguishes between "agreements related to Production Sharing Contracts," which are deemed valid, and "authorizations and memoranda of understanding related to oil and gas," which are deemed invalid.[11] It is something entirely different to speak of "executive, legislative, and judicial authorities of the Iraqi Kurdistan Region [being] considered as legitimate as from the time of their establishment in the year 1992," and then later speak of "[l]egislation enacted in the region of Kurdistan since 1992 . . . remain[ing] in force, and decisions issued by the government of the region of Kurdistan, including court decisions and contracts, [remaining] valid. . . ."[12] The former seems to evidence a clearer intent to break sharply with the earlier formulation; the latter seems more susceptible to being construed as continuing in the same vein, even though undergoing linguistic reconfiguration.

III. CONSTITUTIONAL LIMITATIONS ON KRG ARTICLE 141 POWERS: ARTICLE 23, SECOND, OF THE IRAQI CONSTITUTION AND EXPROPRIATIONS

It is obvious that pursuant to article 141 of the Constitution, the KRG was granted authority to amend or annul all preexisting oil and gas contracts entered into since 1992. However, by virtue of the precise terms

[10] See supra Chpt. 5, Sec. III.

[11] See the contrast between the KRG draft art. 79 and art. 54 of the 2007 KRG oil and gas law (No. 22).

[12] See the contrast between art. 80 of the Kurds' July 28, 2005, draft of the Iraqi Constitution and art. 141 of Iraq's 2005 Constitution.

of article 141, any laws of the KRG that endeavor to amend or annul such contracts must do so in a way that does not contradict the Constitution. An important constitutional limitation that such amendments or nullifications might contradict is article 23, Second, of the Constitution, which protects private property from confiscatory actions of governmental authorities. In pertinent part, article 23, Second, provides that "[e]xpropriation [of private property] is not permissible except for the purposes of public benefit in return for just compensation, and this shall be regulated by law."[13] This is a widespread limitation well known to scholars of the U.S. legal system. Although it does not stop all expropriations – because circumstances exist under which such actions should be executed – it at least constrains expropriation by requiring that when it occurs, the interests of the affected private party must be adequately safeguarded by assuring that the expropriation is for public purpose and accompanied by compensation.

Those familiar with article 23, Second, of the Iraqi Constitution might consider that the protection it accords to private property is available to Iraqis alone, not foreign natural or juridical persons. However, that view is misplaced because the language of the provision fails to connect its protection to a particular nationality; elsewhere in the enunciation of rights and liberties, Iraqis are explicitly named when the Constitution intends to limit those rights and liberties to Iraqis alone. For example, the preceding provision in article 22 references a right to work being vested in all "Iraqis"; article 24 references a freedom of movement among regions guaranteed to all manpower, goods, and capital that is "Iraqi."[14] In this context, the fact that article 23, Second, states nothing about the expropriation limitation being available only to Iraqis suggests that natural or juridical persons of all nationalities may access the protection.

Of more relevance than arguments about the availability of article 23, Second, to Iraqis alone is the suggestion that, if the article's no-expropriation standard were to limit the KRG's exercise of article 141 power to amend or annul preexisting contracts, it would render that power without real consequence. Article 141 authorizes the KRG

[13] See Iraqi Constitution, art. 23, Second, available at www.export.gov/iraq/pdf/iraqi_constitution.pdf (accessed July 15, 2006) and www.krg.org/articles/detail.asp?lngnr=12&smap=04030000&rnr=107&anr=12329 (accessed Apr. 24, 2009).

[14] See id. at arts. 22 and 24.

to amend or annul preexisting contractual arrangements. However, if the article 23, Second, standard relative to expropriations were read to encompass contractual amendments or annulments, it would frustrate the authorization provided in article 141. The fundamental flaw with this argument, however, is that it fails to distinguish between expropriations or confiscations of property rights inherent in contractual agreements, as well as amendments or nullifications of contract that may not constitute expropriations or confiscations addressed by article 23, Second. In other words, if the article 23, Second, prohibition of expropriations, except for public purpose and accompanied by the payment of just compensation, requires confiscations or takings of property rights, and amendments or nullifications of contract do not rise to that level, then it may be less of a limitation on the KRG's article 141 power than it may appear.

It would seem that the appropriate view of the article 23, Second, relation to article 141 should recognize that, depending on the facts of each case, an amendment to or nullification of a preexisting contract may or may not constitute an expropriation. Such a determination could be complicated, with the calculation giving greatest weight to at least the way in which the government sought to manifest the amendment or nullification (e.g., regulation of, versus actual physical interference with, the contract); the extent or degree to which the government's action interfered with reasonable anticipations associated with the contract (e.g., did the action result in only some or virtually all of the contract's anticipated value being lost); and the strength of the nexus between the particular measure taken by the government and promotion of the interest the government claims to be protecting by the amendment or nullification (e.g., is the particular action selected by the government to accomplish the identified interest viewed as more or less likely to meet that goal).

A determination that an amendment to or nullification of a preexisting contract rises to the level of an expropriation presents two further questions. First, was the amendment to or nullification of a contract one that secured a legitimate public benefit? The language of article 23, Second, expressly provides that expropriations are not constitutionally permissible "except for the purposes of public benefit." It is not sufficient for there to be a close nexus between the measure taken by the government and any interest the government may choose

to put forward. Therefore, a reworking or nullification of an Eng.&PC or TSA (see Chapter 5) must reflect on whether such was serving a genuine public benefit. If the concept of "public benefit" were to be seen as synonymous with the idea of "public purpose" so familiar to common-law lawyers, then an amendment or nullification emerging from an improvement in the KRG's financial positioning vis-à-vis the parties involved in the oil- and gas-development activity would not by itself be sufficient. Critical shortages of supply or changes in international-policy positions could meet the constitutional requisite, but certainly not the mere enhancement of one's initial commercial reward because of changed circumstances that supplement negotiating leverage that was previously lacking. If "public benefit" is perceived as meaning significantly less than "public purpose," however, then arguments might surface that suggest any benefit secured by the public as a consequence of an amendment or nullification would legitimate confiscatory activity – including benefits that flow from an enhanced commercial reward.

A second question is whether the impairment of property inherent in the contract affected by the action said to constitute an expropriation or confiscation was actually offset by the provision of compensation. The language of article 23, Second, specifically requires that to be permissible, an expropriation must be accompanied by the payment of "just compensation." It is not enough that the government has an important and legitimate public benefit motivating its confiscatory action. Without compensation that is sufficient in amount, useful in form, and timely in payment, even a reworking or nullification of the hypothesized Eng.&PC or TSA for legitimate public purpose would be constitutionally deficient. Article 23, Second, protects against confiscatory actions of governmental authority, and the article 141 provision of power to the KRG to amend or annul preexisting contracts clearly remains subject to that protection.

It should be noted that the nature of the process to be accorded one facing any potential expropriation is described by article 23, Second, as simply "regulated by law." Thus, just as whether the article 23 notion of "public benefit" coincides with the familiar notion of "public purpose," it may be asked whether "regulated by law" carries the same implications as "due process." Admittedly, the latter notion is not self-defining and its substance and content have only emerged over time

as a result of the various considerations thought to be of invaluable significance in the nation-states that relied on it. To state, as in article 23, Second, of the Iraqi Constitution, that expropriations – even when for legitimate public benefit and accompanied by the payment of just compensation – must simply be under procedures "regulated by law" provides significant latitude regarding how the procedures are to be structured. That is, many of the indicia that prove to be hallmarks of a due-process proceeding may be totally absent in the context of an Iraqi expropriation proceeding; yet, such could be considered as fully satisfying all the necessary requisites demanded by article 23.

Perhaps only time will tell which content is to be attributed to the article 23, Second, references to "public benefit" and "regulated by law." However, before leaving this discussion of the role of article 23, Second, in limiting the ability of the KRG to exercise its amendment and nullification powers under article 54 of the oil and gas law (No. 22), comments regarding how general international commercial law has treated mandated deviations from or breaches of contractual commitments are warranted. Concerning article 54 of the KRG oil and gas law (No. 22), there can be no doubt that it provides a legal directive to review and change PSC-related agreements, and to invalidate all authorizations and MOUs related to oil and gas. The consequence, therefore, is to mandate deviations from or breaches of oil and gas contractual commitments entered into by the KRG since 1992.

Without attempting more than a cursory reference to general international commercial law's treatment of such deviations, one real-life illustration – strikingly similar to the hypothetical situation concerning a preexisting Eng.&PC or TSA with the KRG and presented in Chapter 5 – involves the Iran–U.S. Claims Tribunal decision in the 1989 case of *Phillips Petroleum Co. v. Iran.*[15] In that case of more than twenty years ago, a decade after the revolution in 1979, the Iranian government mandated through legislation a departure from and noncompliance with a contractual commitment between Phillips Petroleum and the Iranian National Oil Company. In this respect, the case parallels what might be expected in the event that a foreign entity, partnered with the KRG in an earlier oil and gas arrangement, could present by

[15] See [Iran–US] *Phillips Petroleum Co. v. Iran* (1989) 21 Iran–U.S.C.T.R. 79 [Iran–US C.T.].

way of a legal claim for any Kurdish exercise of article 54 authority so as to change or back away from its legal commitment. Reviewing the breach of contract for consistency with international legal obligations, the Phillips Petroleum Co. Tribunal concluded that by impairing contractual rights, the action of the revolutionary Iranian government constituted a confiscatory measure in contravention of the dictates of international law. Specifically, the Tribunal indicated that "expropriations by or attributable to a State of the property of an alien gives rise under international law to liability for compensation, and this is so whether the property is tangible, such as real estate or a factory, or intangible, such as the *contract rights involved in the present case*."[16] In the Tribunal's estimation, this conclusion was ineluctable, given both the fact that the language of the specific treaty of comity between Iran and the United States implicated by the case, and the general principles of international law regarding host-state takings of foreign-owned property, mandated avoidance of either direct or indirect confiscations of foreign-owned property, unless for a public purpose – and even then only when accompanied by the prompt payment of just, adequate, and effective compensation.[17]

Especially instructive in the Tribunal's opinion was the expressly stated position on the matter of Iranian legal enactments leaving no choice but to interfere with Phillips's contractual commitment. The Tribunal indicated that: "Changes of such a character and magnitude [as those resulting from the 1979 revolution] are not without consequence to contractual relationships, but they do not affect the validity of . . . [earlier] agreements. In other words, a revolutionary regime may not simply excuse itself from legal obligations by changing governmental policies, nor take for public benefit without compensation businesses operated by foreign private persons under the previous regime."[18] Given the reluctance of the Tribunal to accept the contention that the government adopted legislative directive mandating the involved noncompliance by the Iranian National Oil Company did not insulate it

[16] See id. at para. 76 (emphasis added).

[17] See, generally, "Indirect Expropriations" and the "Right to Regulate" in International Investment Law, OECD Working Papers on International Investment Law No. 2004/4 (Sept. 2004), available at www.oecd.org/dataoecd/22/54/33776546.pdf (accessed Sept. 17, 2009).

[18] See id. para. 86, at 111.

from the legal proceeding, there is reason to expect suspicion to be cast on KRG reworkings or nullifications pursuant to article 54.

IV. CONSTITUTIONAL LIMITATIONS INHERENT IN ARTICLES 15, 23, FIRST, AND 26

Three other provisions of the Iraqi Constitution merit attention in the context of possible limitations on the KRG's ability to exercise power to review and amend, or to nullify and invalidate, preexisting oil and gas contractual undertakings – power defined by article 141 of that same Constitution. Article 15 of the Iraqi Constitution provides that "[e]very individual has the right to enjoy life, security and liberty" and that "[d]eprivation or restriction of these rights is prohibited except in accordance with the law and based on a decision issued by a competent judicial authority."[19] Article 23, First, a companion provision to the article 23, Second, no-expropriation rule, states that "[p]rivate property is protected" and that "[t]he owner shall have the right to benefit, exploit and dispose of private property within the limits of the law."[20] The final provision, article 26, obligates the state to "guarantee the encouragement of investment in the various sectors" of the economy.[21] As discussed herein, however, although all three of these other provisions suffer from their own problems, article 26 proves of most substance relative to KRG actions involving oil and gas contracts entered into since August 1992.

With respect to article 15 of the Iraqi Constitution, it resonates of the due-process provision of the U.S. Constitution. Life and liberty are guaranteed to every individual, and their deprivation is strictly prohibited unless in a manner prescribed by law. It is interesting that rather than also referring to the U.S. Constitution's guarantee of "property," article 15 of the Iraqi Constitution opts for guaranteeing the "security" of all individuals. This difference is not inconsequential, because without the constitutional protection of "property," it is impossible to contend that the prohibition against "deprivation . . . except in

[19] See Iraqi Constitution, supra note 13 at art. 15.
[20] See id. at art. 23, First.
[21] See id. at art. 26.

accordance with the law"[22] provides any protection to contractual arrangements analogous to those in the due-process provision of the U.S. Constitution.

Yet, beyond the deficiency involving the article 15 lack of reference to "property," there is another problem in that the only process that deprivation of guaranteed rights is subjected to by that article is process "in accordance with the law and based on a decision issued by a competent judicial authority."[23] This is a far cry from requiring a deprivation in the context of "due process," with all the attendant procedural and substantive protections which have come accompany that constitutional requisite. Even if property had been protected by the language of article 15 of the Iraqi Constitution, any deprivation incident to a KRG exercise of its authority under article 141 of the Iraqi Constitution, and article 54 of the oil and gas law (No. 22), could have been effected without demanding anything other than the procedures followed be prescribed in the law authorizing them and based on a judicial decision by a court vested with jurisdiction. As distinct from the notion of "due process," the deprivation standard established by article 15 of the Iraqi Constitution is minimal, to say the least.

Article 23, First, the companion to the expropriation standard of article 23, Second, is no more consequential than article 15 for limiting the KRG's contract-altering or contract-invalidating authority. As with article 23, Second, the declarations that private property is protected by the Constitution, and that the owner of property has the right to benefit from and exploit such, are clearly not limited to Iraqis alone. Apparently, these protections also may be invoked by foreign natural or juridical persons. Nevertheless, both declarations of article 23, First, are problematic. To begin with, it is one thing to assert that the Iraqi Constitution protects a private property owner's ability to benefit from and exploit that property, and something quite distinct for the KRG – in dealing with a preexisting agreement related to a PSC – to make modifications after subjecting such an agreement to review by the Regional Council. Modifications in that context could be argued to leave intact sufficient property rights said to be represented by the concerned agreement so as to allow compliance with the constitutional

[22] See supra note 18.
[23] See id.

directive regarding property owners' rights to witness benefit from and exploit property they own. Only in regard to preexisting authorizations and MOUs related to oil and gas would the KRG's designation as null and void by law raise the potential for noncompliance with that constitutional directive. The designation in article 54 of the KRG oil and gas law (No. 22) that such legal arrangements are deemed invalid unless the Regional Council determines otherwise prevents appropriate contract beneficiaries from receiving benefit from or exploiting their property rights.

But even in a situation that involves an authorization or MOU related to oil and gas deemed null and void by the terms of article 54, there are two additional problems presented by the exact language of article 23, First, of the Iraqi Constitution. Initially, the language of that provision provides only for the protection of private property and the owner's right to benefit from and exploit it. Not until the following paragraph of the article 23, Second, no-expropriation provision is a constitutional prohibition on governmental power asserted to secure the article 23, First, protection. What this means is that article 23, First, cannot be looked to – independent of article 23, Second – as an effective limitation on the KRG's power to interfere with preexisting oil and gas contracts. Additionally, the very language of article 23, First, states that the protected rights of private-property owners to benefit from and exploit their property must exist "within the limits of the law."[24] Admittedly, there may be some question about how such language is to be construed (and, clearly, it is to be construed as constraining how owners may claim to be able to use their property). However, it is not wholly unreasonable to suggest that this is a roundabout way of the Constitution taking into account that other provisions (e.g., article 141) provide for derogations of sorts from what might be viewed as constitutional protections of an otherwise absolute nature. Assuming this to have some merit, article 141 is cast in terms more than adequate to provide a "stamp of approval" for the KRG's adoption of an article 54–like legislative measure that alters preexisting contract rights. As a consequence, the article 23, First, reference to "within the limits of the law" seriously undercuts any claim that it limits the KRG's authority to change oil and gas contractual arrangements entered into since 1992.

[24] See supra note 19.

We now consider the language of article 26 of the Iraqi Constitution; language that presents its own problems but is capable of making a claim somewhat greater than articles 15 and 23, First, to limiting the KRG's article 141 and article 54 authority. The language of article 26 provides for a state guarantee regarding the encouragement of investment in the various sectors of the Iraqi economy, whether it is the economy of the Kurdistan Region or elsewhere. As with several of the other provisions examined previously, it cannot be denied that article 26 contains additional language – immediately following its declaration of state guarantee – that references the relationship between the guarantee and other regulatory legal measures. The specific reference provides that the guarantee of encouragement of investment "shall be regulated by law."[25] But rather than sounding in the nature of an opportunity for undercutting the guarantee provided by article 26 – as suggested previously regarding the article 23, First, reference to "within the limits of the law" conditioning that article's protection of private property – the reference is probably best seen as meaning only to signify that the way the state can fulfill the constitutional directive to guarantee encouragement of investment in the various sectors of the economy is to be reflected in the terms and provisions of various legal measures. Regarding the fundamental essence of article 26, the constitutionally declared state guarantee of encouragement of investment admittedly does not directly address either preexisting oil and gas contracts or actions that might be taken to impair them. Nonetheless, there is no doubt that the KRG exercise of authority under article 141 of the Iraqi Constitution, and article 54 of the oil and gas law (No. 22), to modify or terminate contractual arrangements negotiated between 1992 and August 2007 would be counter to the article 26 directive to guarantee the encouragement of investment. Surely, any interference with hard-negotiated legal commitments jeopardizes the interest of affected contracting parties in future investment, thereby conflicting with the article 26 mandate that encouragement of investment be guaranteed by the state.

To be sure, however, as with many other provisions of the Iraqi Constitution, the article 26 directive must be construed in conjunction with seemingly contradictory directives found elsewhere in the same

[25] See supra note 20.

document. That is to say, the article 26 directive must be understood as being in an analogous sort of balance with the authority granted to the KRG by article 141 of the Constitution, as that which exists with the no-expropriation standard of article 23, Second: one directive validates preexisting contracts of the KRG, yet empowers the Kurds to amend or nullify them; the other constrains that empowerment by requiring any expropriation of property to satisfy certain requisites. The article 26 declaration of state guarantee of the encouragement of investment plainly does not share the puissance of the article 23, Second, directive regarding expropriation. In affecting various measures the state may adopt that ripple through the community of economic investors, however, it is not completely lacking in consequence. Whatever consequence it can bring to the matter of handling preexisting KRG oil and gas legal arrangements, it exists in a harmonious balance with the KRG's article 141 and article 54 authority, not in a position of being able to trump that authority.

V. CONCLUSION

From this discussion, the weightiest constitutional limitation on the KRG's ability to review and alter those PSC-related agreements, or nullify and invalidate those authorizations and MOUs related to oil and gas entered into since 1992 derives from article 23, Second, and its expropriation standard. Other provisions in the Iraqi Constitution could be proffered as limiting the KRG article 141 and article 54 powers, but all lack the force of article 23, Second.

At least two concerns merit observation before examining in Chapter 7 the specific types of oil and gas assets that might be affected by legal actions prosecuted by claimants seeking to recover on debts owed by Iraq. One observation has to do with the degree of seriousness evidenced in the Iraqi Constitution regarding the extent to which all laws and legal measures adopted by subcentral governmental units (e.g., the KRG) must comport with the terms of the federal Constitution. The other observation builds on the previous reference to the Phillips Petroleum case[26] and is concerned with the general sense in the international legal community that states are not entirely free with

[26] See text accompanying supra notes 14–17.

complete impunity to impair preexisting contractual commitments with foreign private parties.

With respect to the first observation – separate from the very language in article 141 indicating that any KRG exercise of that article's power must be consistent with the Constitution – there are no fewer than four instances where it is made explicit that the provisions of the Constitution override any measures not consistent therewith. These include article 2, First, paragraph C, which declares that "[n]o law may be enacted that contradicts the rights and basic freedoms stipulated in this Constitution."[27] The private-property rights specified in article 23 are among the relevant rights and freedoms. Paralleling article 2, First, paragraph C, is article 13, which provides in 13, First, that the Constitution "is the preeminent and supreme law in Iraq and shall be binding in all parts of Iraq without exception," and in 13, Second, that "[n]o law that contradicts this Constitution shall be enacted" and "[a]ny text in any regional constitutions or any other legal texts that contradicts this Constitution shall be considered void."[28] Article 46 follows these declarations of supremacy by adding that "[r]estricting or limiting the practice of any of the rights or liberties stipulated in this Constitution [including the rights concerning the sanctity of private property] is prohibited. . . ." Although the article continues by allowing derogations from such rights or liberties by law, it conditions all such laws on the requirement that the "limitation or restriction . . . not violate the essence of the right or freedom."[29] And finally, article 120 of the Constitution, in addressing the powers of the regions, states that the regions cannot try to "bootstrap" their way to superiority over the Iraqi Constitution by adopting their own constitutions and claiming that they authorize or require actions inconsistent with the federal Constitution. Its terms provide that whereas each region shall adopt a constitution of its own, it is with the proviso "that it does not contradict this Constitution."[30] From these four provisions, there is little room for suggesting that the KRG is somehow entitled to employ authorities of articles 141 and 54 in a way that escapes the limitations dictated in other provisions of the Iraqi Constitution.

[27] See Iraqi Constitution, supra note 13 at art. 2, First, para. C.
[28] See id. at art. 13.
[29] See id. at art. 46.
[30] See id. at art. 120.

As for the general sense in the international legal community regarding the matter of government action impairing preexisting contractual arrangements with foreign private entities, there are a couple of particular sources that lean forward on the development of such law and thereby act to propel its further evolution and widespread acceptance. It must be cautioned that the objective here is to provide only a brief introduction to reinforce what has been offered previously regarding specific provisions of the Iraqi Constitution; in no way is the objective to capture the innumerable nuances and complexities associated with these sources.

The first source is the Energy Charter Treaty (ECT) of 1991,[31] a largely European instrument. The terms of the ECT apply to protect energy-related "investments," which are defined to include, among other things, any "rights conferred by law or contract or by virtue of any licenses and permits."[32] Articles 10 and 13, respectively, of that instrument obligate contracting parties to not "impair by unreasonable or discriminatory measures [the] management, maintenance, use, enjoyment, or disposal" of investments,[33] or "nationalize[], expropriate[], or subject[] [investment] to a measure or measures having [an] effect equivalent to nationalization or expropriation"; unless for purposes of securing a public interest; done in a nondiscriminatory manner; following due process of law; and accompanied by the payment of prompt, adequate, and effective compensation.[34]

A second source is reflected in the terms of what had been put forward by the Organization for Economic Cooperation and Development (OECD) in the late 1990s as the proposed Multilateral Agreement on Investment (MAI).[35] The MAI protects "investments," which are defined as including "rights under contracts, including... production or revenue-sharing contracts,"[36] as well as "rights conferred pursuant to... contract[s] such as concessions,

[31] See the Energy Charter Treaty (1991), available at www.ena.lt/pdfai/Treaty.pdf (accessed Apr. 29, 2009).

[32] See id. at art. 1(6)(f).

[33] See id. at art. 10(1).

[34] See id. at art. 13(1).

[35] See Draft Multilateral Agreement on Investment, 22 Apr. 1998, OECD Doc. No. DAFFE/MAI(98)7/REV1, available at www1.oecd.org/daf/mai/pdf/ng/ng987r1e.pdf (accessed Apr. 30, 2009).

[36] See id. at II(2)(iv).

licenses, authorizations, and permits."[37] Tracking the ECT, the MAI's protection for investments extends under Part IV, 1.1, to "full and constant protection and security" from the host state.[38] Part IV, 1.2, indicates that this certainly means the host "shall not impair by [unreasonable and/or discriminatory] measures the operation, management, use, enjoyment or disposal" of investments.[39] Part IV, 2, provides that the host state "shall not expropriate or nationalize directly or indirectly an investment . . . or take any measure or measures having equivalent effect"; unless to secure a public interest; done in a nondiscriminatory manner; under procedures representing due process of law; and with the action being accompanied by the payment of prompt, adequate, and effective compensation.[40]

In addition to the forward-leaning pressure of the ECT and the MAI, it bears mentioning that within the United States, the 1960 Harvard Draft Convention on the International Responsibility of States for Injuries to Aliens, and the more recent 1987 Restatement Third of the Law of Foreign Relations of the United States both offer insight about the development of the thinking of those in the international legal community when it comes to governmental responsibility to avoid actions that impair preexisting contractual commitments.

The Harvard Draft Convention[41] provides at article 10(1) that a "taking, under the authority of the State, of any property of an alien, or of the use thereof is wrongful." This is followed in article 10(3)(a) by the indication that a taking of property "includes not only any outright taking . . . but also any such unreasonable interference with the use, enjoyment, or disposal of property as to justify an inference that the owner thereof will not be able to use, enjoy, or dispose of the property within a reasonable period of time after the inception of the interference."[42] Although the language of article 10 does not explicitly refer to contract rights as a protected form of property, article 12 of the Draft Convention removes any question about their protected nature.

[37] See id. at II(2)(vii).

[38] See id. at Pt. IV, 1.1.

[39] See id. at Pt. IV, 1.2.

[40] See id. at Pt. IV, 2.1.

[41] See Extracts from Draft No. 11 of Convention on the International Responsibility of States for Injuries to Aliens, *1960 Proceedings of the Am. Soc'y of Int'l Law* (Apr. 28–30, 1960) at 102.

[42] See id. at 103, arts. 10(1) and 10(3)(a).

Article 12(1) declares that "[t]he annulment by a State of a contract or concession [between the State and an alien] or nonperformance by the State of the terms of such a contract or concession" is wrongful "if the annulment or nonperformance is inconsistent with the law of the State as it existed at the time of the making of the contract or concession and is effected with the purpose of securing to the State or to other persons... [the] benefits owed to the alien...."[43] Regarding contracts or concessions between an alien and an entity other than the central government, article 12(4) provides that the "annulment or modification by a State... of any [such] contract or concession" is wrongful if it amounts to a "discriminatory departure" from the law of the state concerned, or "constitutes an unreasonable departure from the principles of law which are generally recognized by municipal legal systems as applicable to such contracts or concessions."[44]

Similarly, the Restatement Third provides in section 711 for state responsibility "for injury to a national of another state caused by an official act or omission" that violates "a right to property or another economic interest that, under international law, a state is obligated to respect for persons, natural or juridical, of foreign nationality as provided in §712."[45] Section 712 declares that not only are states responsible for takings of property that are not for public purpose, are discriminatory, or are not accompanied by just compensation,[46] but also for any "repudiation of breach of... contract" where "discriminatory" or "motivated by noncommercial considerations"[47]; where "the foreign national is not given an adequate forum to determine his claim of repudiation or breach"[48]; or where there occurs "other arbitrary or discriminatory acts or omissions that impair property or other economic interests...."[49]

The basic thrust of all these efforts that purport to either recapitulate or contribute to the development of international law addressing the matter of the sanctity of contractual commitments with foreign

43 See id. at 104–05, art. 12(1)(a).
44 See id. at art. 12(4)(a)-(b).
45 See Restatement (Third) of the Foreign Relations Law of the United States, Sec. 711 (1987).
46 See id. at Sec. 712(1).
47 See id. at Sec. 712(2)(a).
48 See id. at Sec. 712(2)(b).
49 See id. at Sec. 712(3).

natural or juridical persons is clearly respect for consensual arrangements and the property interests held by others. The differences among these four texts could present a variety of issues in the context of their application to any situation involving the review and modification of a preexisting KRG agreement related to a PSC, or the nullification and invalidation of a preexisting KRG authorization or MOU related to oil and gas. However, the reason each is raised at this juncture is simply to stress that the international legal community shares deeply in the essential position explored previously regarding the limits imposed by the Iraqi Constitution on the ability of the KRG to exercise its powers under article 141 of the Constitution, and article 54 of the oil and gas law (No. 22) to affect oil and gas commitments negotiated between 1992 and August 2007.

7 OIL AND GAS ASSETS AFFECTED BY THE PROSECUTION OF LEGAL CLAIMS AIMED AT RECOVERING ON DEBTS OWED BY IRAQ

I. INTRODUCTION

As discussed in Chapter 6, oil and gas contractual arrangements can be categorized according to whether they were entered into before or after 1992, whether before or after the removal of Saddam Hussein from power in Spring 2003, or before or after the transfer of governmental power in June 2004 by the CPA to indigenous Iraqi elements. It is clear from the language of article 141 of the Iraqi Constitution and article 54 of the 2007 KRG oil and gas law (No. 22) that the timing of any such arrangements can be consequential. When it comes to the oil and gas assets that a prosecuting claimant may seek in order to secure satisfaction for debts considered to be owed by Iraq, they also may fall into several distinct categories. Generally speaking, although not based on date, those categories include specific oil and gas properties subject to negotiated contract rights, in-kind oil and gas actually lifted from Iraqi fields and in the hands of Iraqis or third parties, and revenues generated by the sale of Iraqi oil and gas and held by Iraqi or third-party entities or institutions.

Rendering each of these three categories more concrete, the specific oil and gas properties subject to negotiated contract rights can involve either of two situations: (1) actual Iraqi-based and exclusively Iraqi-controlled oil and gas fields (the usual situation for all undeveloped and some developed oil and gas reserves); or (2) Iraqi-based oil and

gas fields in which a contractual development or production arrangement exists and can support a foreign collaborator arguing that any PSC, production or profit-sharing agreement, or general oil- and gas-development agreement with Iraq not only accords the foreign collaborator a form of property interest in the fields, but one that is embodied in the particular contract or agreement itself. In contradistinction, the category of in-kind oil and gas actually lifted from Iraqi fields and in the hands of Iraqis or third parties entails oil and gas that already has been extracted from the ground and held in storage facilities, pipelines, or transport vessels. Plainly, in view of the ability of such oil and gas to be moved around – like any contract or agreement said to embody a property interest in Iraqi oil and gas fields and held by a foreign company participating with the Iraqi government in developing its hydrocarbon resources – oil and gas in storage facilities, pipelines, or transport vessels may be found outside as well as inside the boundaries of Iraq. With respect to that category involving revenues generated by the sale of Iraqi oil and gas and held by Iraqi or third-party institutions, they also may be found either in Iraq or outside its borders, and represent earnings on the disposition of oil and gas that originated in Iraq.

This chapter examines each of the three broad categories. It does so, however, by separating the actual assets that an Iraqi creditor might seek in a debt-recovery action from the nature of the legal action brought against those assets, and the person or entity that actually holds them. The recounting of the three categories in this introduction leaves the actual assets, the legal action brought, and the matter of who holds the assets thoroughly entangled; discussion in the following sections of this chapter disentangles them. At the outset, the focus is on the question of how the relevant legal measures that provide protection against legal action for Iraqi oil, gas, and revenues treat various types of Iraqi oil and gas assets. Only then is attention turned to the nature of the legal action brought and the implications deriving from assets held by a transferee or third party rather than an Iraqi entity or institution. Given that Chapter 8 addresses the issue of differences associated with prosecuting a claim for debt repayment in one nation as opposed to another, the comments and analysis herein are limited to what the relevant UN Security Council resolutions establish as the international norm relative to particular assets and holders, as well as

actions prosecuted. The survey begins by examining how the relevant Security Council resolutions regard petroleum, petroleum products, and natural gas. Next, it takes up several matters concerning how the resolutions treat proceeds from the sale of petroleum, petroleum products, and natural gas. Thereafter, attention is focused on actual oil and gas properties, and any legal interests in or shares of such properties or petroleum, petroleum products, natural gas, or proceeds from their sale. Beyond that, the nature of the legal action one might seek to initiate, and whether the person or entity against which it is initiated is a transferee or third party vested with an interest in Iraqi oil and gas or proceeds from their sale or Iraqi oil and gas properties, is examined in detail. Finally, the chapter integrates the concepts emerging from the preceding examination with illustrative judicial decisions involving legal actions to recover on debts owed by foreign governments or their instrumentalities.

II. RELEVANT SECURITY COUNCIL RESOLUTIONS AND OIL AND GAS LIFTED FROM IRAQI FIELDS

Reflecting on the discussion in Chapter 3 about the development of the UN Security Council resolutions concerning the protection of Iraqi oil, gas, and revenues from claims by creditors, recall that after the embargo was terminated and Iraq was again permitted to export oil and gas under the oil-for-food program, resolution 986 of 1995 provided in paragraph 14 that Iraqi "petroleum and petroleum products" were immune from legal proceedings as long as title remained in Iraq.[1] The inclusion of "natural gas" on the list of in-kind protected items entitled to immunity did not surface until its addition in paragraph 22 of resolution 1483 in 2003.[2] In paragraph 27 of resolution 1546 in 2004, the immunity was lost with respect to debts related to final judgments "arising out of a contractual obligation entered into by Iraq after [the] 30 June 2004" handover of governmental authority from the CPA.[3] Resolution 1905, the current authoritative statement by the Security Council, as discussed herein, continues the forcefulness of

1 See supra Chpt. 3, Sec. II.
2 See supra Chpt. 3, Sec. III.
3 See supra Chpt. 3, Sec. IV.

paragraph 22's insulation from legal action, subject to the exception concerning debts incurred after June 2004.[4]

Iraqi petroleum, petroleum products, and natural gas may be immune from legal action until title passes to an initial purchaser, but what about the meaning of petroleum, petroleum products, and natural gas? While few would dispute that it means bulk or barreled loads of crude oil originating in Iraq, does it also include that entire spectrum of separated, distilled, or refined product that derives from the production of crude or natural gas? It may cover crude or natural gas shipped by pipeline, truck, rail, maritime vessel, or otherwise; it also may cover combustible or lubrication oils and gasoline, aviation, or other motor fuels refined from crude. But precisely how far does the definition extend? Does the protection accorded petroleum, petroleum products, and natural gas provide immunity from legal action to products of natural gas? After all, the only reference to the products of hydrocarbons is "petroleum products." For natural gas, no comparable reference is made to "natural gas products." Furthermore, even in terms of petroleum products, how far along the product process does the UN-provided immunity reach? Does it extend beyond combustible or lubrication oils, gasoline, aviation, and motor fuels to all refined items that are the product of Iraqi treatment and processing? Though Iraq is unlikely to be engaged at present in the process of producing these items, what about highly refined household lubricants, distillates, and additives of which crude oil is their base feedstock? Does the protection also extend to any item that is made from petroleum, even though when completed the source of that product itself is no longer recognizable as having its source in crude oil? Again, without any knowledge of whether Iraq produces them, would cloth or cloth-like fibers, tires, gloves, hoses, and other items made from petroleum qualify for immunity as a "petroleum product"?

The fundamental objective of UN-provided immunity for Iraqi petroleum, petroleum products, and natural gas was to ensure the availability of financial resources to rebuild and stabilize that nation. The absence of any clean set of definitions provided in the relevant Security Council resolutions is unfortunate and complicating. Practically speaking, however, the financial resources so essential to Iraq have

[4] See supra Chpt. 3, Sec. V.

long derived from crude-oil exports, which suggests the unimportance of tackling the definitional problems associated with ascribing content to concepts like "petroleum products"; after all, it is only crude that had long been exported. Nevertheless, to the extent that Iraq is involved in producing refined and highly refined petroleum products, or items that are otherwise made from petroleum, the question remains of the coverage of these items by the language of the controlling Security Council resolutions. A couple of points are important in that regard.

First, as noted, the rebuilding and stabilization of Iraq depends in large part on the availability of the requisite financial resources. It may be true that ethnic and sectarian tensions, the ambitions of neighboring states, and the interference of international terrorists are critical determinants in Iraq's immediate and long-term future. However, without the necessary financial resources, the country cannot function and the security situation would rapidly deteriorate. When the relevant UN resolutions were drafted, the focus may have been on financial resources derived from crude-oil exports, because refined and highly refined products and items made from petroleum likely contributed almost nothing to Iraq's foreign-currency earnings. Little, if any, attention may have been given by the drafters to the fiscal potential incident to the wide and unlimited variety of products that could proceed from sophisticated uses and scientific manipulation of petroleum. Despite this, to the extent that they are capable of generating supplements to the nation's financial resource pool, it certainly makes sense to interpret the Security Council resolutions as rendering all such products eligible for protection from claims by creditors. Refined and highly refined petroleum products, as well as items that science is capable of producing from crude, can provide valuable supplements to the needed revenue potential of Iraqi petroleum.

Second, in reaching to not only petroleum but also "petroleum products," the language of the Security Council resolutions is sufficiently broad to extend immunity to the full range of petroleum-derived products. Nothing in the terms of paragraph 22 of resolution 1483, which is confirmed and reiterated by the currently controlling statement of the Security Council (i.e., resolution 1905), suggests that the only products of petroleum considered subject to immunity are those that either still retain most of the essential and typifying

characteristics of crude oil, or represent the first step removed from petroleum itself. The relevant language references both petroleum and "petroleum products" as being subject to immunity, and the common understanding of the latter includes all items produced or derived from petroleum. Although undoubtedly what was intended when the selected phraseology of the resolution was settled on envisioned items such as gasoline, kerosene, lubricant, and fuel oil, the use of the words *petroleum products* left the potential for a more extensive range of items to be included. Given the fact that the resolution also provided immunity to the "proceeds" from sales of petroleum and petroleum products – and in view of the critical importance of such proceeds to the success of Iraq's reconstruction and future stability – there is every reason to interpret "petroleum products" as expansively as possible. If the drafters intended anything else, they certainly would have offered a confined and circumscribed definition of that concept, or at least limited it to those products resulting most immediately from the processing, refining, or transforming of raw petroleum.

Both the importance of maximizing oil revenues for Iraqi reconstruction, and the breadth of the concept "petroleum products," can support the interpretation of the relevant Security Council language as extending immunity from legal action to not only crude oil and refined or highly refined products, but also to items that can be made from petroleum. However, a substantially weighty practical reason suggests that hesitancy is the watchword when an entity avails itself of the opportunity of immunity with regard to items made from petroleum. While the language of "petroleum products" can be interpreted as encompassing all items made or derived from crude oil, and the fiscal demands of Iraqi reconstruction may indicate the importance of maximizing revenues by expanding the ambit of UN-established immunity from legal action, the practical difficulties and administrative problems associated with policing which revenues from the sales of which products are entitled to protection from claimants, because the products were made or derived from petroleum, would be unworkable and nightmarish. A significant array of items may fall into the category of items produced from petroleum and, as a consequence, the revenues from their sale would be eligible for protection. From a practical and administrative perspective, however, it is far more straightforward

and convenient to designate revenues generated from sales of crude oil and the refined and highly refined products produced from it as immune from legal action, and ignore those from sales of items that retain none of the characteristics of items derived from petroleum or petroleum product derivatives. After all, it is unlikely that Iraq will soon be engaged in large-scale production of such items; even if it were, who among us is sufficiently educated about the full range of items made from petroleum or petroleum products to state authoritatively that some sale revenues are protected and others are not? Furthermore, how can we ever be certain from one day to the next that the innovations and creativity of science will not permit the molecular structure of oil to be manipulated and transformed into a new and hitherto never-before-produced wonder of admiration?

Concerning "natural gas," it is imperative to recognize that the Security Council resolutions do not speak of "natural-gas products," but they do speak of both "petroleum" and "petroleum products." Paragraph 22 of resolution 1483, as confirmed and reiterated by resolution 1905, and its immediate predecessor resolution 1859, extends immunity only to petroleum, petroleum products, and natural gas, not to the products of natural gas. In light of the drafters' explicit signification that immunity extends to the products of petroleum, their failure to specify immunity for natural-gas products must be perceived as deliberate. In accordance with the previous explanation, UN-conferred immunity applies not only to crude oil, but also to products refined from crude oil (e.g., gasoline, aviation, and heating fuel). Whereas the plain language of "petroleum products" suggests that all products produced from crude oil (e.g., tires, fibers, and additives) also could be eligible for UN protection from legal action, practicality suggests otherwise. Concerning natural gas, however, whereas the gas in its original state, in its separated or refined state (e.g., methane, butane, and propane), or converted from either state to facilitate transport or storage (e.g., liquefied natural gas) would be immune from legal action, products that have been produced from natural gas or refined or converted gas (e.g., ammonia fertilizer, plastics, dyes, and medicines) would not. Again, the relevant language adopted by the Security Council references only "natural gas" as entitled to protection from legal claims; nothing is mentioned about extending protection to "natural-gas products."

III. REVENUES FROM SALES OF IRAQI OIL AND GAS

As mentioned previously in connection with the second reason for construing the reference to "petroleum products" as encompassing products that are both refined or processed and transformed or derived from crude oil, even though they retain no similarity to crude, the language of relevant Security Council resolutions speaks of "proceeds" from the sale of petroleum, petroleum products, and natural gas as also being under UN-supplied immunity. Specifically, paragraph 22 of Security Council resolution 1483 provides that "petroleum, petroleum products, and natural gas originating in Iraq shall be immune, until title passes to the initial purchaser from legal proceedings against them . . . and that proceeds and obligations arising from sales thereof . . . shall enjoy privileges and immunities equivalent to those enjoyed by the United Nations."[5] Three aspects of this language merit particular comment: (1) whether the reference to "proceeds" includes only the revenues generated by the sale of petroleum, petroleum products, and natural gas but not other forms of payment; (2) the fact that the reference to "proceeds" is supplemented with the additional reference "and obligations," with no suggestion as to what this means; and (3) the fact that proceeds and obligations resulting from sales of petroleum, petroleum products, and natural gas are extended the same "privileges and immunities" as the UN, whereas petroleum, petroleum products, and natural gas are simply said to be immune from legal proceedings.

With respect to the matter of whether the reference to "proceeds" includes more than just monies generated by the sale of petroleum, petroleum products, and natural gas, it cannot be ignored that the common understanding of the term *proceeds* suggests that any remuneration produced by, or any form of payment incident to the sale of Iraqi petroleum, petroleum products, and natural gas would be entitled to the protective effect of the resolution's grant of immunity. There can be no doubt that the common form of payment is a deposit of cash connected with the presentation of letters of credit and other significant shipping documents. At the same time, however, the use of the term *proceeds* plainly indicates that any form of payment proceeding

[5] See supra Chpt. 3, note 21.

from the transaction is squarely within the ambit of the resolution's immunity protection – whether in the form of cash, an accounting credit, a commensurate adjustment for a debt otherwise owed the purchaser by the seller of the relevant oil and gas (i.e., a credit against an account receivable), an exchange of other goods or services in a barter transaction, or any other form of payment for the sale of petroleum, petroleum products, or natural gas. If the payment proceeds from a sale of petroleum, petroleum products, or natural gas, then it represents the "proceeds" of that sale.

Regarding the reference to "and obligations" in the context of the paragraph 22 reference to "proceeds and obligations arising from sales [of petroleum, petroleum products, and natural gas] . . . enjoy[ing] privileges and immunities equivalent to those enjoyed by the United Nations,"[6] it is interesting to recall the earlier discussion in Chapter 3 about the various draft resolutions that preceded the final text of paragraph 22 of resolution 1483. As discussed therein, the May 9 draft provided in paragraph 15 for the so-called Iraqi Assistance Fund to enjoy the "privileges and immunities of the United Nations" and paragraph 21 indicated that "petroleum, petroleum products and natural gas . . . and proceeds of sales thereof, shall be immune from judicial, administrative, arbitration or any other proceedings (including any prejudgment or postjudgment attachment, garnishment, or execution or other action to satisfy judgment) arising in relation to claims, of whatever kind and whenever accrued, against Iraq or any instrumentality or agent thereof. . . ."[7] The May 15 draft provided in paragraph 19 the same language as in paragraph 21 of the May 9 draft, with the exception of inserting after the reference to "proceeds of sales thereof" the additional words "including indebtedness incurred in connection with such sales."[8] Further significance related to the language of these two drafts is addressed herein, but it bears noting that neither draft contained any reference to immunity, or privileges and immunities equivalent to those of the UN, extending beyond proceeds to "obligations" as well. Concerning immunity, both paragraph 21 of the May 9 draft and paragraph 19 of the May 15 draft referenced only petroleum, petroleum products, and natural gas, as well as "proceeds" of their sale, as entitled to immunity. Clearly, then, in paragraph 22 of resolution 1483 (i.e.,

[6] See id.
[7] See supra Chpt. 3, note 33.
[8] See supra Chpt. 3, note 34.

the penultimate version of the two earlier draft resolutions) an addition was made through the inclusion of the language "and obligations" that had not been contemplated in the earlier drafts of the Security Council. The effect of the additional language was to provide legal protection to not only proceeds from oil and gas sales but also to commitments, liabilities, or other legal responsibilities and duties arising from an oil and gas sales contract. A claimant seeking to access Iraqi oil and gas by proceeding against revenues from their sale, or other obligations said to arise from contracts to sell such, would be restricted by the resolution's grant of privileges and immunities equivalent to those of the UN.

As referenced, both the May 9 and May 15 draft resolutions spoke of the proceeds from the sale of petroleum, petroleum products, and natural gas being immune from legal action.[9] Paragraph 22 of resolution 1483, on the other hand, distinguishes between "petroleum, petroleum products, and natural gas" being "immune" from legal action, and "proceeds and obligations" arising from sales of such being accorded "privileges and immunities equivalent to those enjoyed by the United Nations."[10] There are two interesting dimensions associated with this distinction in language. The first is concerned with the idea that "proceeds and obligations" are extended the "privileges and immunities" of the UN, and "petroleum, petroleum products, and natural gas" are made "immune" from legal action. The second concerns the fact that the draft resolutions of both May 9 and May 15 spoke of petroleum, petroleum products, natural gas, and proceeds of their sale being "immune" from judicial, administrative, arbitration, or any other proceedings; while paragraph 22 of resolution 1483 differentiates between immunity for petroleum, petroleum products, and natural gas, and privileges and immunities for the proceeds of their sale.

Regarding the second of these two interesting dimensions, both the May 9 and May 15 drafts indicated that privileges and immunities of the UN were to be extended to the Iraqi Assistance Fund – the draft predecessor of what became in resolution 1483 the Development Fund for Iraq (DFI). At the same time, proceeds from the sale of petroleum, petroleum products, and natural gas were made immune from judicial, administrative, arbitration, or any other proceedings. Given that the

[9] See supra notes 6 and 7.
[10] See supra note 5.

terms of resolution 1483 indicate that proceeds from the sale of oil and gas were to be held by the DFI, it made sense for paragraph 22 of that resolution to provide that both the DFI and the proceeds it would manage should be extended the same privileges and immunities accorded the UN. As a consequence, it is reasonable to conclude that the change from the draft resolutions' formulation of proceeds being "immune" to such proceeds being entitled to "privileges and immunities" equivalent to the UN signified no more than an attempt to align the protection accorded the proceeds of oil and gas sales with the protection accorded the entity charged with managing and administering them.

With respect the dimension of the technical distinction between according proceeds of the sale of oil and gas the "privileges and immunities" of the UN, and making petroleum, petroleum products, and natural gas "immune" from legal action, it bears noting that the exact language of paragraph 22 of resolution 1483 speaks of petroleum, petroleum products, and natural gas that originate in Iraq as being immune from "legal proceedings against them and not . . . subject to any form of attachment, garnishment, or execution."[11] Thus, in divining the precise reach of the paragraph's grant of immunity to petroleum, petroleum products, and natural gas, it may be best gauged by considering the reference to the words *immunity* from "legal proceedings." As rational as that seems, recall also that both the May 9 and May 15 drafts of what became resolution 1483 spoke of immunity from "judicial, administrative, arbitration, or any other proceedings (including any prejudgment or postjudgment attachment, garnishment, or execution or other action to satisfy judgment) arising in relation to claims, of whatever kind and whenever accrued, against Iraq or any instrumentality or agents thereof. . . ."[12] Although it can be suggested that the language of the drafts is more extensive and inclusive than what finally appeared in paragraph 22 of the operative resolution, it cannot be denied that the concept of immunity from "legal proceedings" is sufficiently open-ended to accommodate any proceeding designed to determine legal rights, whether the proceeding be judicial, administrative, arbitral, or of some other form. Neither can it be denied that

11 See id.
12 See supra notes 6 and 7.

the language of paragraph 22 also references petroleum, petroleum products, and natural gas not being subject to "any form" of attachment, garnishment, or execution. Furthermore, recall that paragraph 14 of the Security Council's immediately relevant predecessor resolution (i.e., resolution 986) used precisely the same type of linguistic formulation about petroleum, petroleum products, and natural gas being "immune from legal proceedings and not subject to any form of attachment, garnishment or execution."[13] As a consequence, it is difficult to draw any conclusive and unavoidable inference about a desire to narrow or reduce the scope of immunity from the fact that the more detailed and descriptive language of the May 9 and May 15 drafts was not used.

There are two final points about paragraph 22 of Security Council resolution 1483. The first concerns the import of the use of language providing that proceeds and obligations arising from oil and gas sales (and the DFI) are entitled to "privileges and immunities" equivalent to those enjoyed by the UN. In other words, what does it mean to state that proceeds and obligations arising from the sale of oil and gas (and the DFI) receive the same privileges and immunities as the UN? As previously indicated, nothing should be inferred from the mere fact of the paragraph 22 reference to petroleum, petroleum products, and natural gas being "immune" and proceeds being accorded the "privileges and immunities" of the UN, because both the May 9 and May 15 drafts placed proceeds, as well as petroleum, petroleum products, and natural gas, under the single rubric of "immun[ity]." Nonetheless, are there any consequences of making proceeds (and obligations) of the sale of oil and gas eligible for the same privileges and immunities of the UN?

On that particular score, it appears that the protection of the concept of "privileges and immunities" of the UN reaches substantially farther than the immunity accorded by resolution 1483 to Iraqi-origin oil and gas. It is true that the UN Charter does not explicitly define the privileges and immunities to which that international organization is entitled. Article 105(1) of the Charter merely provides that "[t]he Organization shall enjoy in the territory of each of its Members such privileges and immunities as are necessary for the fulfillment of its

[13] See supra Chpt. 3, notes 9–16.

purposes."[14] Shortly after the adoption of the UN Charter, however, the nation-states represented in the UN entered into the Convention on the Privileges and Immunities of the United Nations (February 13, 1946), which specified at length the nature of UN rights and protections.[15] In addition to providing the UN with complete immunity from any and all legal process, article II, section 3 of that Convention clearly states that all property and assets of the organization are totally immune from any form of "executive, administrative, judicial, or legislative action."[16] Essentially, then, while immunity from legal process – as provided by resolution 1483 to Iraqi petroleum, petroleum products, and natural gas – may be of a substantial nature, the immunity granted to the UN by the terms of the Convention on Privileges and Immunities extends as well to immunity from actions of every imaginable branch of government. The language of paragraph 22 in resolution 1483, therefore, must be read as according protection to the proceeds of oil and gas sales – as well as obligations arising from their sale – that is more extensive than that which is provided to Iraqi oil and gas proper. The proceeds and obligations of oil and gas sales are immune from executive, administrative, judicial, or legislative action, whereas oil and gas as such are immune only from actions to determine legal rights, whether in the form of judicial, administrative, arbitral, or other action taken, but not action that is executive or legislative in form.[17]

[14] See UN Charter, art. 105(1).

[15] See Convention on the Privileges and Immunities of the United Nations, 13 Feb. 1946, 1 U.N.T.S. 15.

[16] See id. at art. II, Sec. 3.

[17] It is appreciated that a potential difficulty with this argument is that even action that is executive or legislative in form can be understood as capable of determining legal rights and, therefore, seen as legal action from which resolution 1483 declares petroleum, petroleum products, and natural gas immune. What makes such a suggestion difficult to accept is the fact that paragraph 22 of resolution 1483 explicitly states that petroleum, petroleum products, and natural gas are to be "immune" from "legal action," and proceeds and obligations arising from sales thereof (as well as the DFI) are to receive the "privileges and immunities" that are accorded "the United Nations." As suggested herein, the mere fact that "proceeds" in the draft resolutions had been treated the same way as petroleum, petroleum products, and natural gas and was then changed should not, standing alone, be taken to mean that proceeds are to receive a different form of protection. However, the fact that the same paragraph references both the concepts of "immun[ity]" and "privileges and immunities" of the UN, with the latter having particular meaning in international law, certainly suggests that the two concepts do not have the same meaning.

The second of the two final points relates to the argument just made – in that it seems to provide corroboration for the notion that "immun[ity]" means something different from UN-like "privileges and immunities" – and it also has to do with the effect worked on paragraph 22 of resolution 1483 by the language of paragraph 27 of Security Council resolution 1546. As observed in Section II of this chapter and Section IV of Chapter 3, paragraph 27 of resolution 1546 was adopted in anticipation of the transition of governmental authority from the CPA to the local Iraqis, and it provided for the removal from paragraph 22 of "privileges and immunities . . . with respect to any final judgement [sic] arising out of a contractual obligation entered into by Iraq after 30 June 2004."[18] By confining the removal of protection to "privileges and immunities," the resolution appeared to confirm that a distinction exists between the "privileges and immunities" accorded proceeds and obligations arising from oil and gas sales (and the DFI), and the "immun[ity]" provided petroleum, petroleum products, and natural gas. More important for this study, however, the language of paragraph 27 subjected the proceeds and obligations of oil and gas sales (and the DFI) to the full range of executive, judicial, administrative, and legislative action. This must be kept in mind when considering the proceeds or revenues generated by the sale of Iraqi oil and gas. Those proceeds or revenues can be accessed to satisfy a final judgment connected with an Iraqi contractual obligation entered into after the end of June 2004.

IV. WHERE PARAGRAPH 27 OF RESOLUTION 1546 FITS INTO THE MIX OF SECURITY COUNCIL RESOLUTIONS PROVIDING PROTECTION TO IRAQI OIL AND GAS ASSETS

As indicated previously, paragraph 27 of resolution 1546 removes the protection provided to the proceeds of Iraqi oil and gas sales in connection with "any final judgment arising out of a contractual obligation entered into by Iraq after 30 June 2004."[19] The consequences of this paragraph extend well beyond corroborating the distinction between

[18] See text accompanying supra note 3.
[19] See id.

"immun[ity]" and "privileges and immunities" spoken of in paragraph 22 of resolution 1483 and making the point that because Iraqi oil and gas proceeds were once protected from legal action does not mean they remain forever protected. Paragraph 27 of resolution 1546 is important for the other questions it raises regarding legal action (see Chapter 4, Section III). One question is concerned with how it is possible to have a paragraph 27 final judgment for which the privileges and immunities of paragraph 22 in resolution 1483 were removed if those privileges and immunities are in force (by the terms of the resolution 1905 extension of paragraph 22) until at least December 31, 2010?[20] Another question concerns what paragraph 27 means in the reference to the removal of privileges and immunities regarding "any final judgement [sic]"[21]? A third question concerns the significance of the paragraph's indication that the final judgment of relevance must be one resulting from a "contractual obligation"[22] that Iraq entered into after the end of June 2004.

Indeed, addressing each question seriatim, it is strange that paragraph 27 speaks of the removal of privileges and immunities from legal action in the event of a final judgment when an essential thrust of resolution 1483 was to prevent actions that might eventuate in a final judgment. Fundamentally, however, this misunderstands the relationship between resolutions 1546 and 1483. Paragraph 22 of resolution 1483, extended and reaffirmed by resolution 1859 and then 1905, provided "immun[ity]" from legal action for Iraqi petroleum, petroleum products, and natural gas. The same paragraph also accorded UN-like "privileges and immunities" to proceeds and obligations arising from the sale of oil and gas (and the DFI). With adoption of paragraph 27 of resolution 1546 (reaffirmed and endorsed by resolution 1905, and its predecessor, 1859), the privileges and immunities that were earlier extended to oil and gas proceeds and obligations (and the DFI) were removed in the context of actions concerning Iraqi contractual obligations after June 30, 2004, and resulting in a final judgment for which satisfaction is sought. In other words, Iraqi petroleum, petroleum products, and natural gas cannot be the object of a legal action, but actions based on a contractual obligation of Iraq entered into after the end of

[20] See text accompanying supra Chpt. 3, Sec. V.
[21] See supra note 18.
[22] See id.

June 2004 that eventuate in a final judgment can serve as the basis for seeking satisfaction from oil and gas proceeds and obligations (and the DFI). The idea of paragraph 27 is to put Iraq in the same position as all other nations relative to post–June 30, 2004, contractual obligations, at least with regard to the proceeds from the sale of oil and gas and the obligations arising from them. That is, if the Iraqi government incurs a contractual obligation, it cannot insulate proceeds from the sale of oil and gas from creditor action.

Before addressing the meaning of paragraph 27 of resolution 1546 in reference to "any final judgement [sic],"[23] it is interesting that the language of that paragraph can be interpreted in a manner distinct from what has been suggested. Specifically, it can be maintained that paragraph 22 of resolution 1483 stopped legal actions against Iraqi petroleum, petroleum products, and natural gas, and also provided privileges and immunities to oil and gas proceeds and obligations (and the DFI). And when paragraph 27 of resolution 1546 removed the insulation of privileges and immunities regarding final judgments resulting from Iraqi post–June 2004 contractual obligations, it indicated that Iraq and all its instrumentalities (as distinct from oil and gas proceeds and obligations and the DFI) had never been protected from actions of any form. Support for this view is in the fact that the May 9 and May 15 drafts of what became resolution 1483 provided for insulation of the Iraqi government and its instrumentalities and organs,[24] but paragraph 22 of resolution 1483 fails to contain that actual language. Thus, although an action against Iraqi petroleum, petroleum products, natural gas, or the related proceeds and obligations (or the DFI) could not be prosecuted under paragraph 22, actions could be taken against the Iraqi government or its organs. As a consequence, paragraph 27 of resolution 1546 signified that a final judgment against the government or its organs based on a contractual commitment entered into after June 30, 2004, could be used to access the proceeds from the sale of oil or gas and associated obligations (or the DFI).

Of course, the fatal flaw in this interpretation of the relationship between resolutions 1483 and 1546 is the fact that actions of Iraqi creditors would not be prosecuted against Iraqi petroleum, petroleum

23 See id.

24 See text accompanying supra notes 7 and 8.

products, natural gas, or the proceeds or obligations associated with the sale of oil and gas but rather against the Iraqi government or one of its instrumentalities. It is through these actions that creditors would access Iraq's oil, gas, or proceeds. And given that a basic aim of resolution 1483 was to free up that oil, gas, and proceeds for the reconstruction and stabilization of Iraq, the insulation of the Iraqi government or its instrumentalities from causes of action was essential. With the adoption of paragraph 27 of Security Council resolution 1546, the world community signified only that it was willing to subject the proceeds from and obligations associated with the sale of oil and gas to use for the satisfaction of final judgments arising from Iraq's contractual obligations entered into after June 30, 2004. It was not signifying recognition that final judgments against the Iraqi government or its instrumentalities or organs existed because legal action against that government or its instrumentalities had long been permitted, and that those judgments – if based on a contractual obligation after the end of June 2004 – could serve as the foundation for accessing oil and gas proceeds and obligations related thereto.

Now as to the meaning of "any final judgement [sic]" in paragraph 27, it should be recalled that the language permits actions to recover against oil and gas proceeds and obligations related to those sales for Iraqi failure to fulfill contractual obligations entered into after June 2004. At first glance, it could be thought the notion of a final judgment refers to one rendered by a judicial body of the highest appellate level, or below the highest level if the decision of a lower judicial body was not appealed. Stated more directly, the notion of a final judgment requires a penultimate decision of a court of law. There is no doubt that this understanding of the notion of a final judgment is reasonable, yet it is incomplete and far too limited. The better interpretation of "any final judgment"[25] is that which includes not only judgments rendered by courts of law but also those that are final and resulting from nonjudicial proceedings. Nothing in the terminology of paragraph 27 of Security Council resolution 1546 dictates that the final judgment of relevance be rendered by a judicial body. Indeed, quite to the contrary, the language of paragraph 27 merely requires that there exist "any"[26] final judgment.

[25] See supra note 18.
[26] See id.

Furthermore, the language of paragraph 22 of resolution 1483 on which resolution 1546 operates simply speaks of "legal proceedings"[27] in the context of granting immunity to Iraqi oil, gas, proceeds, and obligations arising from sales (and the DFI). Therefore, not only does it seem that the basic immunity accorded by the appropriate Security Council resolutions relates to immunity from legal proceedings of any sort – whether pursued before a judicial, arbitral, administrative, or other body – the removal of that immunity with respect to "any final judgement [sic]"[28] concerning Iraqi contractual obligations entered into after June 30, 2004 also relates to final judgments rendered by a judicial, arbitral, administrative, or other body that determines legal rights and responsibilities.

Regarding the last of the questions posed earlier – that is, the significance of the reference in paragraph 27 of resolution 1546 to final judgments involving a "contractual obligation"[29] – it might be asked how that serves to inform or limit the reach of the resolution's removal of privileges and immunities. It is obvious that concerning the alterations made in the various earlier drafts considered by the Security Council (see Chapter 3, Section IV), the language of "contractual obligation" excludes final judgments based on tort, criminal, regulatory, or other noncontractual matters.[30] Only final judgments concerning issues of contractual obligation are eligible to benefit from the paragraph 27 removal of the privileges and immunities accorded the proceeds of and obligations arising from the sale of Iraqi petroleum, petroleum products, and natural gas. Additionally, the matter of a contract giving rise to the final judgment need not involve a contract concerning Iraqi oil and gas; all that is required by paragraph 27 to be eligible for the removal of privileges and immunities is a "contractual obligation" that gives rise to a related final judgment. Finally, the contractual matter that results in the relevant final judgment must be one for which the existence of an "obligation" was demonstrated. It is not enough to secure a final judgment in the context of raising a mere claim of contractual obligation. To meet the full requisites of paragraph 27, the final judgment

[27] See supra note 2.
[28] See supra note 18.
[29] See id.
[30] See supra Chpt. 3, Sec. IV.

must arise from an "obligation"[31] based in contract. A final judgment grounded on a noncontractual obligation – even though rendered in the context of an unsuccessful assertion of contractual obligation – fails to meet the standards of paragraph 27 in resolution 1546.

V. ACTUAL IRAQI OIL AND GAS PROPERTIES, AND LEGAL INTERESTS IN OR SHARES OF SUCH PROPERTIES OR OIL AND GAS AND REVENUES FROM SALES

In addition to the matters previously examined in this chapter, there is another that concerns actual oil and gas properties, or the shares in such properties or oil and gas and revenues from their sale. The language of paragraph 22 of Security Council resolution 1483 – as modified by paragraph 27 of resolution 1546 and reaffirmed and extended by resolution 1905 – insulates Iraqi oil and gas from legal action and grants UN-like privileges and immunities to proceeds from their sale. What about the actual properties from which the oil and gas have been produced or lifted? What about fractional-share interests in such properties, the oil and gas produced there from, or the revenues generated by their sale? Do the terms of the relevant Security Council resolutions protect the oil, gas, and proceeds, but not the properties that generated them, the share interests in the properties, the oil and gas they have produced, or the revenues resulting from their sale?

Before answering these questions, brief comment should be offered on a transitional matter related to the discussion in the preceding section – that is, the fact that Security Council resolution 1483, in granting privileges and immunities equivalent to those of the UN to the proceeds of the sale of Iraqi oil and gas, provides that the same privileges and immunities also are granted to "obligations"[32] arising from those sales. The reference to "obligations"[33] being entitled to privileges and immunities is important in that it insulates from any form of action all those obligatory legal commitments associated with the sale of Iraqi oil and gas. This means insulation from any form of action instituted against the beneficial interest of a purchaser who has contractually bound the

[31] See id.
[32] See supra Chpt. 3, note 21.
[33] See id.

Iraqi government or one of its instrumentalities to deliver oil and gas. The language of paragraph 22 of resolution 1483 provides that not only are Iraqi oil and gas immune from legal action, but "obligations"[34] arising from their sale are also entitled to UN-like privileges and immunities. Similarly, any other form of "obligation"[35] resulting from the sale of oil and gas, including those unrelated to the actual delivery of oil and gas, is insulated from judicial, arbitral, administrative, and other forms of action.

Now to the central matters of whether oil and gas properties – or fractional-share interests in them, their production, or the proceeds of that production – are protected under the terms of resolution 1483. It cannot be denied that paragraph 22 speaks of oil and gas as well as proceeds and obligations being entitled to insulation from forms of action. However, it makes little sense to provide for such insulation and then permit the prosecution of actions against the very properties responsible for producing the oil and gas and the associated proceeds and obligations. If a central aim of paragraph 22 of resolution 1483 is protection of Iraqi oil and gas and revenues from claims by creditors in order to ensure necessary financial resources for the country's reconstruction and future stability, then it is incongruous to permit actions to be brought against the actual oil and gas properties responsible for that production and revenue. This is equally valid whether the concern is with actions against the entire or a fractional-share interest in the properties. If none of the oil and gas produced in Iraq is subject to action by creditors until title has been transferred to a purchaser, then why would it be reasonable to permit action against a fractional-share interest in an Iraqi oil and gas property capable of producing oil and gas? If the oil and gas produced from that fractional-share interest is insulated from creditor action, then the interest from which it can be produced also should be.

Redirecting the focus to whether insulation is also provided for share interests in the oil and gas actually produced or the proceeds generated by their sale, it is clear that the plain language of paragraph 22 of resolution 1483 contains nothing along those lines. The language of the paragraph is completely silent about protection for

[34] See id.
[35] See id.

entire and fractional-share interests; it only provides that "*petroleum, petroleum products and natural gas* originating in Iraq shall be immune, until title passes to the initial purchaser . . . and *proceeds* and obligations arising from sales thereof . . . shall enjoy the privileges and immunities" enjoyed by the UN.[36] Conceivably, this language is broad enough to encompass interests in the entirety of oil, gas, and proceeds as well as fractional-share interests; it does not distinguish between the two levels of interest. In this connection, however, two points must be emphasized.

First, it is possible that the conveyance or granting of a fractional-share interest in oil and gas could be understood as creating the type of title passage to an initial purchaser that the language of paragraph 22 speaks of as producing a removal of the immunity from legal proceedings. Again, paragraph 22 refers to Iraqi petroleum, petroleum products, and natural gas being immune "until title passes to the initial purchaser."[37] The language obviously contemplates fully executed sales of Iraqi oil and gas to initial purchasers, and it is just as possible that it also extends to executed or completed sales – not of tangible oil and gas but rather fractional-share interests in them. In either transaction, the passage of title would lift the immunity otherwise provided for in resolution 1483.

Second, even with acceptance of the preceding point, it cannot be ignored that the phrasing of paragraph 22 suggests a distinction between any fractional-share interests in oil and gas as such, and fractional-share interests in proceeds or revenues from their sale. To reemphasize, the language provides that oil and gas are immune from legal proceedings, until title passes to the initial purchaser. Regarding the proceeds from the sale of oil and gas and the privileges and immunities that they are accorded, no language is present that indicates that an event resulting in a transfer of title to an initial purchaser lifts the relevant privileges and immunities. The clearest reason for this is the fact that any event or activity creating a transfer of title to the proceeds from the sale of oil and gas means that the revenues from such are no longer proceeds of the Iraqis. To restate the point, it makes no sense to conclude that revenues or proceeds produced by the sale of

36 See supra note 2 and Chpt. 3, note 21 (emphasis added).
37 See id.

Iraqi oil and gas are forever imbued with paragraph 22 privileges and immunities, regardless that those have left Iraqi control, entitlement, or accounts that Iraq may access. In the event that Iraq were to sell oil and gas and spend the proceeds generated, the entity then controlling the proceeds as a consequence of the spending controls nothing to which the privileges and immunities of paragraph 22 extend. The entity into whose hands the proceeds pass has the exclusive legal claim thereto. Not only is it illogical to think a creditor of Iraq could access those proceeds to satisfy a debt owed by Iraq, it also is illogical to think that the entity into whose hands the proceeds have passed could claim a continuing effect of the original privileges and immunities of the UN in order to insulate it from claims of its own creditors.

To summarize, then, oil and gas properties and fractional-share interests in them are nowhere referred to in the operative language of resolution 1483 (reaffirmed and extended by resolution 1905) as eligible for protective insulation from claims. Nevertheless, it certainly makes sense that if oil and gas themselves are insulated from legal actions, then the properties that produce them are also eligible. Concerning fractional-share interests in oil and gas that have been produced or in the proceeds generated by their sale, it is imperative to distinguish between fractional-share interests in oil and gas and fractional-share interests in proceeds. The reason for this resides in the strict language of paragraph 22, which indicates that transactions resulting in title being transferred from Iraq to an initial purchaser open "petroleum, petroleum products, and natural gas originating in Iraq" to claims by creditors, as the immunity otherwise available is lost. Because that language fails to differentiate between transfers of an entire interest in Iraqi oil and gas – as opposed to a fractional-share interest – conceivably even the transfers of the latter could produce the same result. The strict language of paragraph 22 completely leaves aside "proceeds" when it comes to the idea of the grant of immunity being lost. The language concerning "proceeds" states only that they are to enjoy the privileges and immunities of the UN, with no reference to such being lost.[38] Not until the adoption of resolution 1546 and its paragraph 27 was any provision made for proceeds to lose those privileges and immunities, which was to occur only with respect to final judgments

[38] See supra note 2 and Chpt. 3, note 21.

arising from contractual obligations entered into by Iraq after the CPA transfer of governmental authority in June 2004.

VI. NATURE OF THE LEGAL ACTION BROUGHT AND AGAINST WHOM MAINTAINED

From the previous discussion, there is little room for doubt regarding the severe limitations faced by creditors seeking to access Iraqi petroleum, petroleum products, or natural gas, the proceeds and obligations arising from their sale (or the DFI). Before addressing the nature of the action those creditors might be interested in taking against such assets, something should be said about the relationship to the notion of immunity being lost for petroleum, petroleum products, and natural gas when title is transferred to an initial purchaser. Recall that paragraph 22 of UN Security Council resolution 1483 – the operative resolution concerning immunity for Iraqi assets – states that petroleum, petroleum products, and natural gas of Iraqi origin are immune from legal proceedings, until title is transferred to the initial purchaser. The effect of this lifting of immunity on transfer of title basically means that any creditor of the transferee can pursue satisfaction of the debt owed by targeting any petroleum, petroleum products, or natural gas that may have originally been produced in or held by Iraq but subsequently was lawfully transferred to a purchaser. It is not beyond the realm of possibility that ineffective and unsuccessful attempts to effectuate transfers from Iraq to another may occur, and creditors of the transferee may then be frustrated in their efforts to recover against what had attempted to be transferred. In these instances, creditors of Iraq would not be in any better position, for title would remain in Iraq and the basic immunity of the resolution would continue to apply. One thing is clear: A successful transfer of title would not empower a creditor with a claim against Iraq – except, perhaps, in the most exceptional cases – to access the transferred Iraqi assets simply because after the transfer, title is no longer vested in Iraq but rather in an initial purchaser. After all, as in any other analogous situation, it is difficult to imagine a transferee of an asset having that asset subjected to claims pressed by creditors of the transferor, especially if – as long as the title remains in the transferor's hand – the asset is generally immune from creditor

claims. Perhaps there are highly exceptional cases in which transferred assets can be subjected to claims by creditors of Iraq, even after titles have been transferred to initial purchasers. Plainly, such cases would have to involve circumstances that suggest Iraq leveraging the immunity for Saddam-era debts provided by the Security Council in a way that allows it to gain an advantage or benefit not otherwise available.

Leaving this matter aside, and moving to the nature of the cause of action being prosecuted against Iraq, the question is whether it is of any consequence if the action is initiated before a judicial body, a nonjudicial alternative dispute-settlement entity, an executive-branch claim-resolution organ, or another form of noncourt, nonjudicial mechanism. Paragraph 22 of Security Council resolution 1483 references petroleum, petroleum products, and natural gas as being immune from "legal proceedings"[39] and proceeds and obligations arising from their sale as being entitled to privileges and immunities like those enjoyed by the UN. Does that somehow equate to immunity from causes of action initiated before a court of law and only a court of law? Or is the notion of immunity from "legal proceedings" and privileges and immunities of the UN, sufficient to provide insulation from any type of action, whether maintained before a judicial-, executive-, or legislative-branch organ or before a privately constituted and governmentally recognized substitute form of dispute-resolution mechanism?

As was seen in connection with the scope of the Iraqi assets designed to be protected by the language of resolution 1483, the breadth of paragraph 22 appears quite extensive when speaking of the nature of proceedings from which the Iraqi assets are protected. The precise language of that paragraph not only references immunity of oil and gas from "legal proceedings," it also continues by providing "and not be subject to any form of attachment, garnishment, or execution."[40] In other words, whether the action of concern is, strictly speaking, a "legal proceeding[]" or one that would result in an attachment or garnishment of, or execution on, Iraqi petroleum, petroleum products, or natural gas, paragraph 22 precludes it from moving apace regardless of its form. As the language declares, Iraqi-origin petroleum, petroleum products, and natural gas are immune from legal proceedings and not

39 See supra note 2 and Chpt. 3, note 21.
40 See id.

subject to "any form"[41] of attachment, garnishment, or execution. The inclusive nature of the words *any form* of attachment, garnishment, or execution is more than capable of protecting against not only judicially instituted attachment, garnishment, or execution, but also any taking of petroleum, petroleum products, or natural gas, or satisfaction extracted from such, by operation of any other form of nonjudicial or alternative dispute settlement or resolution mechanism.

If the drafters intended paragraph 22 to be confined to court proceedings aimed at attachment, garnishment, or execution, they would have been explicit in stating it. Instead, they opted for the more open-ended idea of immunity from "legal proceedings" designed to effect an attachment, garnishment, or execution in "any form." Conjoining the reality of this situation with the fact that the language used in the earlier referenced May 9 and May 15 draft predecessors to paragraph 22 of resolution 1483 indicated immunity "from judicial, administrative, arbitration, or any other proceedings (including any prejudgment or postjudgment attachment, garnishment, or execution or other action to satisfy judgment),"[42] it is fair to construe the language of paragraph 22 as simply a more parsimonious and succinct way to express the same encompassing and all-inclusive breadth captured in the two drafts. It is not impossible, but it would seem inaccurate, to view the Security Council's move from the broader language of the drafts to that of immunity from "legal proceedings" effecting attachment, garnishment, or execution in "any form" as evidence of intent to strictly limit the immunity to actions before courts of law. Again, this could have been expressed by the Security Council if that had been its intent. In relying on the notions of "legal proceedings" and in "any form," the Council apparently chose to implement the inclusive aims of the two drafts with a slightly different linguistic formulation.

In addition to the nature of the action that might target Iraqi oil, gas, proceeds, and obligations, or oil and gas properties (or the DFI), it should be asked if any significance attaches to whether the entity prosecuted is the government or an instrumentality of Iraq, or a third party with an interest in any of the relevant Iraqi assets. For example, assume a case in which the Iraqi Ministry of Oil (MoO) or one

41 See id.

42 See text accompanying supra notes 7 and 8.

of its various divisions or companies, or the Iraq National Oil Company (INOC), or Iraq's State Oil Marketing Organization (SOMO) has actual possession of oil or gas that was lifted from an Iraqi hydrocarbon field. Assume another case in which the oil or gas found its way into the lawful physical possession of a foreign oil and gas company as payment for technical services provided to Iraq in an effort to develop that hydrocarbon field. Again, the language of paragraph 22 of Security Council resolution 1483 indicates that Iraqi petroleum, petroleum products, and natural gas are immune from legal proceedings until title passes to an initial purchaser, whereas proceeds and obligations arising from their sale (and the DFI) are entitled to privileges and immunities equivalent to those of the UN. Pursuant to paragraph 27 of resolution 1546, the privileges and immunities accorded proceeds and obligations (and the DFI) are lost with respect to final judgments concerning Iraqi contracts entered into after June 30, 2004. In the context of these two examples, these specific international rules mean that the oil and gas held by the MoO (its divisions or companies), INOC, or SOMO would remain immune from creditor action, as long as Iraq retained legal title. Oil and gas in the hands of the foreign company and representing compensation for services rendered, however, would be subject to creditor action. Although it is true that the language of paragraph 22 signifies immunity is lost whenever title is transferred to an initial "purchaser," surely that particular term is not to be taken so literally that it would be viewed as not including the recipient of oil in exchange for the provision of services. *Purchaser* is usually understood to include one who obtains something in return for the payment of money, but there is no reason why payment must take that precise form. Section V in Chapter 8 elaborates on this point in the context of examining complexities of the language of various U.S. Executive Orders that offer insulation from creditor action prosecuted in U.S. courts against Iraqi oil and gas assets.

Another point to observe is that the creditors of relevance in regard to the loss of immunity are those of the foreign oil and gas company – not the Iraqi government – except, perhaps, in a limited number of cases accompanied by highly unusual circumstances that suggest continued susceptibility to claims by Iraq's creditors, even after it appears that the requisites of a formal transfer of title have been satisfied. The thrust of paragraph 22 of resolution 1483 is to provide immunity to

Iraqi petroleum, petroleum products, and natural gas while title to such remains in Iraqi hands. When title is transferred to an initial purchaser, the creditors of that purchaser may be able to access the petroleum, petroleum products, or natural gas. Iraq's creditors, however, could not seek from those transferred assets any satisfaction for debts owed by Iraq. Title to the transferred assets are then held by the initial purchaser and, therefore, generally subject only to claimants of the purchaser. Only in the exceptional situation of Iraq leveraging its internationally granted immunity in a way that permits an advantage or benefit not otherwise available does it seem reasonable to consider permitting Iraq's creditors to access assets to which title was transferred to another. Situations like this undoubtedly are few and far between.

VII. HOW THE PRECEDING INTERFACES WITH ILLUSTRATIVE JUDICIAL DECISIONS INVOLVING ACTIONS TO RECOVER ON DEBTS OWED BY FOREIGN GOVERNMENTS OR THEIR INSTRUMENTALITIES

Keeping in mind the review in the preceding sections of this chapter, it is natural to question the completeness, thoroughness, or comprehensiveness of the immunity provided Iraqi oil and gas (by the relevant UN Security Council resolutions and other provisions of law), especially in light of judicial decisions rendered in U.S. courts that involve attempts to recover on debts owed by foreign governments or their instrumentalities. Many of those decisions permit creditors to move forward on recovering debt owed, despite the fact that immunity is otherwise available to the foreign entities concerned. Along this line, the preface of this study referenced the *Consarc*, *Brown & Root International* and the *Commercial Bank of Kuwait* cases, which concerned actions pressed against the Iraqi government for failure to meet debts owed to creditors.[43] Specifically in regard to *Commercial Bank of Kuwait*,[44] recall that Iraq's Rafidain Bank reportedly suffered a default judgment in U.S. federal court on the basis of letters of credit and other guarantees issued by it and the CBI in the context of Iraq's failure to repay

[43] See supra Preface at notes 15–21.
[44] See supra Preface at notes 20–21.

loans earlier made by the Commercial Bank of Kuwait.[45] The Iraqi failure to repay developed in the wake of the First Gulf War. As suggested in the discussion of the *Commercial Bank of Kuwait* case in the preface, subsequent appellate-court decisions concerning that lower-court decision[46] followed, which provide a modicum of illumination on debtor-government liability for borrowing and other commercial activity when immunity seems to be available.

The lower court's decision in *Commercial Bank of Kuwait*, which granted a requested default judgment against the Iraqi financial organs, was subsequently upheld in a 1994 federal appellate-court decision. However, it is incorrect to interpret the conjunctive effect of these two decisions as suggesting reluctance in U.S. federal courts to view grants of immunity to foreign sovereigns or their instrumentalities as deserving of little significance. While it is true that overhanging the *Commercial Bank of Kuwait* decisions was the immunity extant in the U.S. Foreign Sovereign Immunities Act[47] (FSIA) (which is beyond the scope of this study),[48] the overwhelming majority of cases involving the FSIA hinge on whether its various exceptions and conditions permit the immunities to be successfully invoked. Decisions that suggest exposure to legal action and run against the protective effect of immunity, therefore, turn on the peculiarities of surrounding factual circumstances and their relation to technical requirements of the law, not a foundational reluctance in U.S. courts toward the notion of immunity for foreign sovereigns. In the context of the immunity against legal action provided to Iraq and proclaimed by the UN Security Council in resolution 1483 (and then extended in December 2009 by resolution 1905), this indicates that there is nothing inherent in the *Commercial Bank* case and its progeny that, standing alone, suggests the UN-proclaimed immunity provided Iraqi oil and gas, proceeds from their sale, and the DFI is not likely to be honored. Indeed, the Second Circuit federal appellate court in *Commercial Bank* was more than prepared to accord the Iraqi creditors immunity, except for the fact that the FSIA removed immunity for sovereign acts in many instances, including

[45] See id.

[46] See supra Preface at note 20 citing a 1994 appellate decision involving the same matter as the 1992 federal district court decision.

[47] See 28 U.S. Code Secs. 1330, 1602–1611.

[48] See supra Preface. For another reference to FSIA, see infra Chpt. 9, note 20, and accompanying text.

those in which (as here) acts occurring outside the United States were related to commercial activity that resulted in a direct effect inside the United States. In the estimation of the appellate court in *Commercial Bank*, the facts of the case fell squarely within the statutory language of this exception; thus, immunity was deemed lost.[49] One might be inclined to understand *Commercial Bank's* reluctance to permit Iraq to wrap itself in the available immunity as a reluctance born out of aversion to impeding creditors from proceeding in U.S. courts against foreign sovereign debtors. However, the correct understanding of the case is more properly based on Iraq's inability to avoid certain technical statutory exceptions to the invocation of immunity.

Continuing with this theme, there have been other U.S. federal court cases that one might be inclined to read as evidencing a reluctance to honor immunity doctrines aimed at protecting foreign sovereigns, their organs, and instrumentalities. In that vein is the case of *Af-Cap Inc. v. Chevron Overseas (Congo) Limited*, a 2007 decision of the Ninth Circuit federal court of appeals.[50] Admittedly, *Af-Cap* neither involved the government of Iraq in any way nor did it hold immunity to be unavailable. As in *Commercial Bank of Kuwait*, as well as *Consarc* and *Brown & Root International* cases, *Af-Cap* involved legal action by a creditor to collect on a debt owed by a foreign sovereign. Furthermore, it evidenced a court not willing to immediately endorse the application of immunity and agonizing over whether immunity should be recognized.

Although a large number of parties were involved in the *Af-Cap* case, at the core it concerned the failure of the Republic of Congo to pay on a loan on which the creditor was entitled to collect through a garnishment action executing an earlier judgment in its favor. Somewhat distinct from *Commercial Bank* and other cited cases, the assets sought by the creditor in satisfaction of the debtor's nonpayment were (1) not in the debtor's hands but rather a third party (i.e., Chevron Overseas [Congo] Limited) and legally owed by that third party to the debtor; and (2) assets that represented bonuses, taxes, and royalties owed by the third party to the debtor on oil, gas, and other resource-extraction contracts. The loan creating the Republic of Congo's debt, however, was on an unrelated highway-construction project.

[49] See 15 F3d 238 (2d Cir. 1994).
[50] See 475 F3d 1080 (9th Cir. 2007).

Factually, in the early to mid-1980s, the Republic of Congo secured a loan from Equator Bank, with the funds to be used for a highway-construction project. Thereafter, Congo defaulted on the loan, and Equator Bank assigned its interest to Connecticut Bank of Commerce, which obtained an English judgment against Congo and subsequently converted it into enforceable judgments in the United States. The judgments eventually passed to Af-Cap, Inc., which then commenced a creditor garnishment action. However, the action was brought against various Chevron entities with oil, gas, and natural-resource contracts with Congo but that named the Republic of Congo as well. These contracts legally obligated each entity to pay Congo certain bonus, tax, and royalty amounts associated with resource activity.

Among the central questions facing the Ninth Circuit in the case was the protective effect of the immunity provided by the FSIA. As intimated earlier FSIA typically offers immunity to property of foreign sovereigns. However, the Act allows those sovereigns to waive such, and the facts indicated that the Congo had done so earlier in the original loan agreement with Equator Bank. Any waiver then has the effect of exposing assets considered to be "used for a commercial activity in the United States"[51] and, located therein, as susceptible to legal process. The Ninth Circuit, by agonizing over the availability of immunity and deciding against the action by the creditor to secure satisfaction on the Congo's failure to pay its debt, did no more than determine the narrow question of whether the assets of the Congo held by the third-party Chevron entities were assets "used for a commercial activity in the United States." In finding that the language required the property concerned to be actually "put into action, put into service, availed or employed *for* a commercial activity" and not merely "*in connection* with a commercial activity or *in relation* to a commercial activity,"[52] the Ninth Circuit held that the Congo's assets in Chevron's hands were immune from legal action. The Ninth Circuit's decision should not be perceived as an indirect indication of federal-court hesitancy regarding the application of immunity doctrines. Ultimately, FSIA immunity was granted and the labor evidenced by the court in arriving at that decision relates exclusively to the nuanced complexities of the Act's language,

[51] See FSIA, supra note 47 at Sec. 1610(a).

[52] See *Af-Cap*, supra note 50 at para. 38 of the opinion (emphasis added).

not any reluctance to accord the full range of immunities available to foreign sovereigns.

Given the holding in the *Af-Cap* case, it is impossible to construe it as undercutting, eroding, or casting in a new and less-expansive light the otherwise broad and inclusive nature of the immunities provided by UN Security Council resolutions, and other measures, to Iraqi oil and gas. As will be recalled, though the breadth of the most significant Security Council pronouncement is not unlimited in its protection, it does state: " . . . unless the Council decides otherwise, petroleum, petroleum products, and natural gas originating in Iraq shall be immune, until title passes to the initial purchaser from legal proceedings against them and not be subject to any form of attachment, garnishment, or execution . . . and that proceeds and obligations arising from sales thereof, as well as the Development Fund for Iraq, shall enjoy privileges and immunities equivalent to those enjoyed by the United Nations."[53] As far as it goes, this statement is certainly a rather comprehensive grant of immunity. However, because the *Af-Cap* case involved a factual situation in which the creditor was attempting to access foreign-sovereign debtor assets that had no direct relation to what caused the debt, and because the Ninth Circuit determined that the targeted assets were immune from protection, this could be construed to suggest that assets characterized by linkage to the debt itself would be viewed by U.S. courts as generally not immune from legal action. In regard to Iraqi oil and gas, the consequence of drawing any such inference from the *Af-Cap* case could be seen as exposing such oil and gas, the revenues from their sale, and the DFI to legal action whenever there is an immediate nexus between the Iraqi debtor assets targeted and the contractual commitments with the creditor. This would recognize a distinction – in terms of access to immunity – between a case in which a creditor, for example, has contracted to provide oil-field services to an Iraqi governmental entity that defaults on payment, with Iraqi oil and gas then being targeted to satisfy any emerging legal claim, and a case in which a creditor, for example, has loaned Iraq funds to develop a system of hospitals, with Iraq then defaulting on repayment, and the creditor seeking satisfaction from the proceeds generated by the sale of Iraqi oil and gas.

[53] See text of para. 22, Security Council Resolution 1483, supra Chpt. 3, note 3.

At least two points are relevant in connection with making any such distinction on the basis of the *Af-Cap* case. First, there is no apparent language in the relevant Security Council pronouncement that supports such a distinction. It is true that the pronouncement delineates among "petroleum," "petroleum products," "natural gas," "proceeds and obligations arising from sales thereof," and the "Development Fund for Iraq." This is far different, however, from suggesting that it carves out an exception or exclusion from the immunity typically granted in instances in which a creditor's claim results from a contract, legal commitment, or obligation not tied to the particular Iraqi debtor asset from which satisfaction is sought. Second, the *Af-Cap* case clearly permitted the invocation of immunity in those instances in which the foreign sovereign was able to show that assets the creditor moves against may be in the United States, but are not being used there in commercial activity. However, to construe this as meaning that immunities from legal action are lost (including under the Security Council's Iraq resolutions) when the source of the creditor's legal action does not match the debtor asset moved against is to give the *Af-Cap* case far too distended a reading. The case was limited to determining the availability of the FSIA's immunity; holding it remained available because the debtor assets proceeded against were not used in the United States in commercial activity.

Although it may be accurate to conclude that, since the assets in Chevron's hands moved against by the creditor were bonuses, taxes, and royalties owed by, but not yet paid to, the Republic of Congo, they were not assets being used in commercial activity in the United States. This is far from meaning that for an entity to avail itself of relevant immunities, it must be shown that the asset targeted by the creditor is linked in terms of nature or character to the debtor contract or obligation giving rise to the claim. Applied to Iraqi oil and gas, the *Af-Cap* case simply evidences that the FSIA's immunity obtains in every instance in which foreign sovereign property moved against in the United States is not used therein for commercial activity. If it is used for commercial activity, then the Act's immunity is lost. The *Af-Cap* decision should not be interpreted as suggesting reluctance by federal courts to apply concepts of immunity in instances in which there is no nexus between assets targeted by a creditor and the basis for the legal action against the debtor. Indeed, in the *Af-Cap* case itself,

it can be argued that little such linkage existed. Furthermore, it is likely that had the Ninth Circuit been confronted with a situation in which the Congo was directing the use of its receivables in Chevron's hands for investment and other commercial purposes, the FSIA's immunity would have been seen as lost. This would have permitted the creditor to successfully move against the assets, not because they were not linked to the foundational basis of the debtor's obligation, but simply because of the idiosyncrasies of FSIA's statutory language. Regarding immunities provided to Iraqi oil and gas by Security Council resolutions and other measures, it thus seems that the key determinant is the language of those resolutions and measures, not whether the *Af-Cap* case directs perception about a needed connection between assets sought by a creditor and contracts or obligations taken on by a debtor.

A couple of additional problems reflective of the matter of completeness, thoroughness, or comprehensiveness of the immunities provided by relevant Security Council resolutions and related measures deserve comment. One problem is concerned with the fact that – as already seen in the *Af-Cap* case – there are at least some conceptions of immunity provisions that envision creditors being able to proceed against third parties in their attempt to satisfy liabilities of a foreign-sovereign debtor. After all, in that case, the creditor would have been successful in its action against the assets held by Chevron (even though to satisfy a debt of the Republic of Congo) had the factual circumstances regarding the assets and commercial activity in the United States been configured differently. In contradistinction, however, Section VI notes that the Iraqi oil and gas immunity contemplated by the relevant Security Council resolutions will support action against those assets only after they have been transferred to an initial purchaser – and then only for debts owed to the claimant by that transferee (as distinct from Iraq or its instrumentalities).[54] In the absence of highly exceptional circumstances that suggest an Iraqi attempt to leverage its immunity to an advantage not otherwise available, assets in the hands of a third-party transferee are not open to legal action to satisfy the debt of the original Iraqi transferor.

The other problem warranting comment concerns a twist on the problem just examined – specifically, whether Iraqi oil and gas,

[54] See supra Sec. VI.

although held by the Iraqi government, its organs, or instrumentalities, receives the protection of the relevant Security Council immunity resolutions and other measures when targeted by a creditor seeking satisfaction of a debt owed by a private party. For all intents and purposes, such a situation would more than likely involve a case in which a creditor is owed in-kind oil and gas, revenues generated by their sale, or a fractional-share interest in an oil and gas property in return for services performed for a private third-party debtor (foreign or Iraqi) that contracted to provide the Iraqi government with certain oil- and gas-development activities. The private third-party debtor may have nothing other than a claim against a share of the oil and gas produced, revenues from their sale, or oil and gas properties – title to which still may be held in the hands of the Iraqi government. As a consequence, the so-called creditor that provided the private third party with the services forming the basis of the claim may seek recovery against the Iraqi oil and gas (or revenues or properties) to satisfy its own claim. Given the immunities provided Iraqi oil and gas by the Security Council, the viability of such legal action appears predetermined by the suggestion in Section V of this chapter. Recall that when addressing the matter of immunity and fractional-share interests, that section acknowledged the sensibility of providing immunity to such interests, but cautioned that the creation of fractional shares could trigger the "title transfer" language of paragraph 22 of resolution 1483 (as extended by resolution 1905), and that it requires a distinction be drawn between fractional-share interests in oil and gas and fractional-share interests in proceeds from their sale.[55] These considerations would seem to require a role in the calculus of any action by a creditor to collect from assets held by Iraq in satisfaction of debts owed by a third-party private entity.

VIII. CONCLUSION

The language of paragraph 22 of resolution 1483, and paragraph 27 of resolution 1546, (as reaffirmed and extended by Security Council resolution 1859 and the resolution 1905) reveals a wide array of intricacies and complexities associated with the UN protection of Iraqi

[55] See supra Sec. V, especially text accompanying notes 36–38.

petroleum, petroleum products, and natural gas, as well as proceeds and obligations generated by their sale, from claims by creditors. With a broad brush, these provisions insulate Iraqi oil, gas, and proceeds or revenues from their sale from judicial, administrative, arbitral, or other actions that aim to attach, garnish, execute, or otherwise access those assets. It does not matter that a creditor seeks satisfaction from a mere fractional-share interest in Iraqi oil, gas, or proceeds. Neither does it matter that the satisfaction sought is from oil and gas properties, rather than what those properties actually produced. In either case, the protection of the relevant resolutions is applicable. Were creditors permitted to access fractional-share interests or interests in properties distinct from oil, gas, or proceeds, the basic objectives of the Security Council protection would be frustrated.

Once there has been a transfer of title of petroleum, petroleum products, or natural gas from Iraq to an initial purchaser, then a creditor of that purchaser may be able to access what was transferred in order to satisfy any debts owed. Given that any such transfer represents acceptance that the asset involved now belongs to the transferee, creditors of the Iraqi transferor generally cannot proceed against the transferred asset in order to seek satisfaction for the debt owed. Concerning the proceeds or obligations arising from the sale of oil and gas (or concerning the DFI), the privileges and immunities extended by resolution 1483 are lifted relative to contractual obligations, and only contractual obligations, entered into after June 30, 2004. This acknowledges that the transfer of governmental authority from the CPA to the indigenous Iraqi authorities carries with it the responsibilities that flow from negotiated contractual obligations.

A final matter that bears mentioning is what the language of the late-2008 SOFA between the United States and Iraq[56] has to say in relation to the assets insulated from claims by creditors and the nature of the actions against which protection is provided. Article 26 of the SOFA contains three operative paragraphs, with the second and third

[56] See Agreement Between the United States of America and the Republic of Iraq on the Withdrawal of United States Forces from Iraq and the Organization of Their Activities during Their Temporary Presence in Iraq (Nov. 2008), available at http://graphics8nytimes.com/packages/pdf/world/20081119_SOFA_FINAL_AGREED_TEXT.pdf (accessed June 3, 2009).

being the most instructive. Paragraph 1 recognizes the importance of Iraq's oil, gas, and other resources and its financial and economic assets located abroad (including the DFI) to the country's rehabilitation and future development. In light of that, the United States thereby commits to support Iraq in its efforts to secure international debt forgiveness, and achieve a comprehensive and final resolution of all outstanding reparation claims.[57] Paragraph 2 provides that the United States recognizes and understands Iraq's concerns with claims of creditors based on actions of the Saddam-era regime, and that the "President of the United States has exercised his authority to protect from United States judicial process the Development Fund for Iraq and certain other property in which Iraq has an interest." Furthermore, the United States commits to remain engaged with Iraq regarding the continuation of such protections.[58] Paragraph 3 provides that the United States – in full recognition that the extensions of the resolution 1483 protections were scheduled to expire at the end of December 2008 – remains committed to working in the Security Council to assure the extension of the protections "established in Resolution 1483 (2003) and Resolution 1546 (2003) for petroleum, petroleum products, and natural gas originating in Iraq, proceeds and obligations from [the] sale thereof, and the Development Fund for Iraq. . . ."[59] Of course, the adoption of Security Council resolution 1859 in mid-December 2008 effected the extension referenced in paragraph 3, and that has since been followed up by resolution 1905's extension.

Chapter 8 examines in detail the relevant terms of article 26. With respect to the suggestions in paragraphs 2 and 3 of that article, however, it is clear that they address protection for precisely the same assets as the Security Council resolutions discussed in this chapter. Indeed, paragraph 3 of article 26 not only favorably references resolutions 1483 and 1546, it also takes pains to mention petroleum, petroleum products, and natural gas, the proceeds and obligations from their sale, and the DFI by name. It is interesting that when it comes to the nature of the proceedings against which Iraqi assets are to receive protection,

57 See id. at art. 26(1).
58 See id. at art. 26(2).
59 See id. at art. 26(3).

paragraph 2 of article 26 references the notion of "judicial process." That reference, of course, is in the context of the United States acknowledging Iraq's concerns with claims against its assets and the indication that the President has moved to protect those assets from the U.S. judicial process. This stands in clear contrast to the apparently broader scope of protection provided by the Security Council's resolutions.

8 ISSUES ASSOCIATED WITH PROSECUTING A CLAIM IN THE UNITED STATES WITH A GLIMPSE AT OTHER JURISDICTIONS: THE IMPACT OF NATIONAL LAW

I. INTRODUCTION

Security Council resolution 1483 – and others that have provided protection to Iraqi oil and gas and related assets – mandates that the international protection established by the resolution be implemented by member states through the adoption of domestic legal enactments. As indicated in Chapter 7, paragraphs 2 and 3 of article 26 of the SOFA between the United States and Iraq stress the U.S. commitment to protecting Iraqi assets through domestic means, and its commitment to doing the same through the UN. Article 26(2) provides in relevant part that "[r]ecognizing and understanding Iraq's concerns with claims based on actions perpetrated by the former regime, the President of the United States has exercised his authority to protect from United States judicial process the Development Fund for Iraq and certain other property in which Iraq has an interest."[1] Article 26(3) provides, in part, that the "U.S. remains committed to assist Iraq in connection with its request that the United Nations Security Council extend its protections and other arrangements established in Resolution 1483

[1] See art. 26 (2), Agreement Between the United States of America and the Republic of Iraq on the Withdrawal of United States Forces from Iraq and the Organization of Their Activities during Their Temporary Presence in Iraq (Nov. 2008), available at http://graphics8.nytimes.com/packages/pdf/world/20081119_SOFA_FINAL_AGREED_TEXT.pdf (accessed June 3, 2009).

(2003) and Resolution 1546 (2003) for petroleum, petroleum products, and natural gas originating in Iraq, proceeds and obligations from sale thereof, and the Development Fund for Iraq. . . ."[2]

The language of article 26(2) speaks in terms of the fact that the President of the United States "has exercised" his authority to protect from U.S. judicial process both the DFI and certain other property in which Iraq has an interest. This chapter addresses the many issues associated with this protection. Prior to the discussion, however, it must be observed that the significance of the language of paragraph 2 being cast in the past tense relates to various presidential Executive Orders and Notices dating from as early as the months following Saddam Hussein's removal from power in 2003. These Executive Orders and various Notices have kept in place the type of protection from legal action mandated by Security Council resolution 1483 and its relevant successors, including resolutions 1546, 1859, and 1905. The most recent presidential issuance in this line is the Notice of May 20, 2009, entitled Continuation of the National Emergency with Respect to the Stabilization of Iraq.[3] That Notice builds on former President George W. Bush's Notice of May 20, 2008,[4] and extends the insulation from legal action accorded Iraqi oil, gas, and proceeds, and the DFI, until May 22, 2010.

II. DEVELOPMENTAL BACKGROUND OF U.S. PROTECTIVE MEASURES: EXECUTIVE ORDERS 13303 AND 13364

As mentioned previously, in the United States, the seminal domestic measure that implements the UN obligation was adopted in the wake of the removal of Saddam Hussein from power in spring 2003. With greater specificity, that measure was Executive Order 13303,

[2] See id. at art. 26(3).

[3] See Notice of May 20, 2009, Continuation of the National Emergency with Respect to the Stabilization of Iraq, available at www.whitehouse.gov/the_press_office/Message-from-the-President-and-Notice-of-Continuation-regarding-Iraq (accessed June 3, 2009) (hereinafter, Notice of May 20, 2009).

[4] See Notice of May 20, 2008, 73 *Fed. Register* 29683, available at http://regulations/vlex/com/vidpart-v-38400176 (accessed June 3, 2009) (hereinafter Notice of May 20, 2008).

adopted May 22, 2003.[5] It was modified by various other Executive Orders, most importantly Executive Order 13364 of November 29, 2004, which reflects the changes in the UN-established protection regime in Security Council resolution 1546 and the lifting of protection regarding final judgments concerning contractual obligations of Iraq entered into after the end of June 2004.[6]

Executive Order 13303 has several separate sections. Section 1 establishes the basic protections of Iraqi oil and gas assets. It declares that "any attachment, judgment, decree, lien, execution, garnishment, or other judicial process is prohibited, and shall be deemed null and void with respect to: (a) the Development Fund for Iraq, and (b) all Iraqi petroleum, petroleum products, and interests therein, and proceeds, obligations, or any financial instruments of any nature whatsoever arising from or related to the sale or marketing thereof, and interests therein, in which any foreign country or a national thereof has any interest, that are in the United States, or that are or hereafter come within the possession or control of the United States persons."[7] Section 3(c) defines a "United States person" to mean "any United States citizen, permanent resident alien, entity organized under the laws of the United States or any jurisdiction within the United States (including foreign branches), or any person in the United States."[8] Section 3(b) defines "entity" to mean "partnership, association, trust, joint venture, corporation, group, subgroup, or other organization."[9] Section 3(d) defines "Iraqi petroleum, petroleum products" to mean "any petroleum, petroleum products, or natural gas originating in Iraq, including any Iraqi-origin inventories, wherever located."[10] Clearly, the purpose of Executive Order 13303 is implementation of the Security Council directive established by resolution 1483.

With the adoption of Security Council resolution 1546, and its removal of insulation for Iraq from creditor actions targeting proceeds and obligations arising from the sale of Iraqi-origin petroleum,

[5] See Executive Order 13303 (May 22, 2003), available at http://edocket.access.gpo.gov/2003/pdf/03–13412.pdf (accessed June 4, 2009).

[6] See Executive Order 13364 (Nov. 29, 2004), available at http://edocket.access.gpo.gov/2004/pdf/04–26686.pdf (accessed June 4, 2009).

[7] See Executive Order 13303, supra note 5 at Sec. 1.

[8] See id. at Sec. 3(c).

[9] See id. at Sec. 3(b).

[10] See id. at Sec. 3(d).

petroleum products, and natural gas (and the DFI) when based on a final judgment concerning an Iraqi contractual obligation entered into after the end of June 2004, Executive Order 13364 was issued. This Executive Order altered the language of Executive Order 13303 in several significant ways.

First, it inserted language to account for the fact that protection from creditor action was not envisioned by the Security Council after title to petroleum, petroleum products, or natural gas had transferred to an initial purchaser. Thus, section 1 of Executive Order 13364 changed Executive Order 13303 by relabeling section 1(a) and (b) as section 1(a)(i) and (ii), and by inserting in the relabeled section 1(a)(ii) (which references Iraqi petroleum, petroleum products, proceeds, obligations, and interests therein as immune from creditor action) "but only until title passes to the initial purchaser."[11] The effect was to leave the relabeled and amended section 1(a)(ii) stating immunity for "all Iraqi petroleum, petroleum products, and interests therein, but only until title passes to the initial purchaser, and proceeds, obligations, or any financial instruments of any nature whatsoever arising from or related to the sale or marketing thereof, and interests therein, in which any foreign country or a national thereof has any interest, that are in the United States, or that are or hereafter come within the possession or control of United States persons."[12]

Second, Executive Order 13364 added entirely new language to section 1 of Executive Order 13303 to address the situation that much of the revenues generated by the sale of Iraqi oil and gas are held in CBI accounts maintained in the Federal Reserve Bank of New York. That new language of immunity – beyond what was provided to the DFI and Iraqi oil and gas, proceeds, and obligations – is placed in what is now section 1(a)(iii) of Executive Order 13364 and refers to immunity for "any accounts, assets, investments, or any other property of the Central Bank of Iraq, or held, maintained, or otherwise controlled by any financial institution of any kind in the name of, on behalf of, or otherwise by the Central Bank of Iraq."[13]

The third way in which Executive Order 13364 altered the language of Executive Order 13303 is related to the removal in resolution 1546 of

11 See Executive Order 13364, supra note 6 at Sec. 1(a)(ii).
12 See id.
13 See id. at Sec. 1(a)(iii).

protection for creditor claims arising from contractual claims of Iraq following the transfer of power by the CPA. Section 1 of Executive Order 13364 added to the relabeled, amended, and expanded section 1 of Executive Order 13303 a new section 1(b). Essentially, that new language provided that section "1(a) . . . shall not apply with respect to any final judgment arising out of a contractual obligation entered into by the Government of Iraq, including any agency or instrumentality thereof, after June 30, 2004."[14] It is obvious that this tracks the previously addressed language of paragraph 27 of resolution 1546. And obviously, its thrust is to declare that the immunities and protections from causes of action set forth by section 1(a)(i-iii) for the DFI, Iraqi oil, gas, proceeds, obligations, and interests therein, and the CBI, do not apply with respect to final judgments arising from contracts entered into by Iraq after the end of June 2004.

III. PRINCIPAL DIFFERENCES FROM SECURITY COUNCIL RESOLUTIONS 1483 AND 1546

As for the principal differences between the two Executive Orders and Security Council resolutions 1483 and 1546, several comments are in order. At this juncture, however, it must be pointed out that the significance of these differences is not discussed. That discussion would raise questions regarding whether provisions of U.S. domestic law (e.g., the International Emergency Economic Powers Act) can provide authority for certain aspects of the Executive Orders in the event that they are viewed as exceeding what the Security Council resolutions envision. In the following sections of this chapter, the differences between what the Executive Orders call for and the Security Council resolutions envision are categorized and examined from the perspective of what they mean under U.S. law. These differences concern several matters: the nature of the process from which protection is provided; the physical location of the asset from which a creditor might seek satisfaction for the debt owed to it; the significance associated with the relevant asset being one in which an entity other than Iraq or one of its instrumentalities holds an interest; the applicability of immunity in the context of

[14] See id. at Sec. 1(b).

fractional-share interests as opposed to the entirety of the asset; the applicability of immunity to assets other than Iraqi petroleum, petroleum products, natural gas, and proceeds and obligations from their sale (and the DFI); and the inclusion of the CBI in the protection specifically created by Executive Order 13364.

Regarding the nature of the process from which protection is provided, it will be recalled that the relevant Security Council resolutions spoke in terms of "legal proceedings."[15] Although it was suggested that that language is probably best understood as any type of proceeding designed to determine legal rights and interests, especially given the drafting history behind resolution 1483, it is conceivable that some might interpret it as limited to proceedings before the judicial branch of government. The language of the *chapeau* (i.e., opening head note) of the Executive Orders bestowing protection for Iraqi assets is more than explicit on this matter. Again, section 1 of both Executive Orders 13303 and 13364 clearly states that Iraqi assets are not subject to "any attachment, judgment, decree, lien, execution, garnishment, or other judicial process."[16] Although it is accurate to call attention to the fact that the reference to "any attachment, judgment, decree, lien, execution, [or] garnishment" suggests a wide range of actions from which Iraqi assets are protected, it cannot be ignored that the language in the section 1 chapeau ends its catalogue of actions from which protection is available with the words "or any other judicial process." The employment of the term *other* in this context indicates that the reach of the Executive Orders' protection does not extend beyond actions that involve "judicial process." This clearly indicates that the Executive Orders establish protection only from judicial-branch actions. Such a reading of the chapeau would permit administrative, arbitral, and other nonjudicial actions to proceed against Iraqi assets, given that judgments or decisions rendered in all such proceedings are not self-enforcing and depend on subsequent judicial action to render their ultimate and final opinions meaningful. However, there is reason to believe that in going no farther than providing protection from "judicial process," the Executive Orders provide adequate protection for those assets. Nevertheless, it cannot be denied that the Executive Orders' limitation

[15] See supra Chpt. 7, Sec. VI.

[16] See text accompanying supra note 7 for the precise language of Executive Order 13303.

to "judicial process" can result in significant problems and inconveniences for the Iraqi government and its instrumentalities. Arbitral or administrative determinations require subsequent recourse to judicial processes to secure enforcement, but the mere fact that the former may proceed under the Executive Orders can be extremely problematic for those required to defend themselves.

As for the fact that the Executive Orders speak of the concerned Iraqi assets as those "that are in the United States, or that are or hereafter come within the possession or control of United States persons,"[17] nothing in the language of Security Council resolution 1483 or 1546 establishes such a requirement. The most apparent reason that this is lacking from the terms of the UN resolutions is concerned with two specific facts: First, the Security Council was casting resolutions in language sufficiently broad to relate to all member states, not only the United States; and second, it recognized rather implicitly that Iraqi assets to be protected against actions brought by creditors in most instances would be assets physically situated within the territorial borders or the judicial jurisdiction of the nation-state in which the action was prosecuted. In light of these two facts, it is not unusual to find that the U.S. Executive Orders implementing the details of the Security Council's protective scheme provide for protection of Iraqi assets when located in the United States or under the control of a U.S. person.

The language of the Executive Orders references the particular Iraqi assets – whether Iraqi petroleum, petroleum products, natural gas, proceeds and obligations from their sale, interests therein, or other financial instruments arising from the sale of their marketing – as assets "in which any foreign country or a national thereof has any interest."[18] This difference from what is provided for in the Security Council resolutions is striking in adding breadth to the UN insulation against creditor actions. In looking at Security Council resolutions 1483 and 1546, they aim basically at protecting Iraqi assets while in the control of the Iraqi government. As discussed previously, paragraph 22 of resolution 1483 states that Iraqi-origin oil and gas shall be immune – until title passes to the initial purchaser – from a range of legal proceedings. Proceeds and obligations arising from the sale of such (and

[17] See id.
[18] See id.

the DFI) shall enjoy the same privileges and immunities as the UN. The former suggests that transfers of title to oil and gas from Iraqi hands to initial purchasers remove the insulation from claims by creditors (although they then would have to be claims brought by creditors of the transferee). The latter does not expressly link the privileges and immunities to which proceeds and obligations (and the DFI) are entitled to such being in Iraqi hands, but it seems to be implicit. As discussed previously, few would seriously contend that because proceeds since passed from the Iraqi government to another entity by way of a purchase are said to be originally proceeds from the sale of oil and gas, such monies always retain UN-like privileges and immunities. These proceeds retain protection only while in Iraqi hands or under Iraqi control. The language of section 1(b) of Executive Order 13303 and section 1(a)(ii) of Executive Order 13364 stands in stark contrast. In providing for protection from judicial process for "all Iraqi petroleum, petroleum products, and interests therein, and proceeds, obligations, or any financial instrument of any nature whatsoever arising from or related to the sale of marketing thereof, and interests therein, *in which any foreign country or a national thereof has any interest*,"[19] the language not only protects relevant Iraqi assets still in Iraqi hands or under the control of the Iraqi government, it also provides protection for such assets when held or controlled by a vast range of other entities. It is true that Executive Order 13364 added to the quoted language the words "but only until title passes to the initial purchaser." Before that addition, the language of section 1(b) of Executive Order 13303 accorded protection from creditor claims to all Iraqi oil and gas held by, for example, American or other non-Iraqi international companies. With the added language of Executive Order 13364 regarding the passage of title to an initial purchaser, the breadth of the protection was considerably reduced. Nevertheless, Iraqi-origin oil and gas held by the same American or non-Iraqi international company remained protected from judicial process until transferred to an initial purchaser.

With respect to the general reference to "interests therein" that appears in the language of section 1(b) of 13303 and section 1(a)(ii) of 13364, the notion is capable of addressing what is discussed in Chapter 7, Section V, concerning the applicability of protection to

[19] See id. (emphasis added).

mere fractional-share interests in relevant Iraqi assets. As discussed, that matter was far from explicitly addressed in the language of resolutions 1483 and 1546.[20] The Executive Orders providing for immunity under U.S. law, however, approach this matter directly with reference to "interests therein." Specifically, the language provides protection for all appropriate Iraqi assets "and interests therein." As just observed, it establishes protection from judicial proceedings for all Iraqi oil and gas "and interests therein," as long as title has not been transferred to an initial purchaser, and for proceeds, obligations, or financial instruments arising from or related to the sale or marketing thereof, "and interests therein."[21] Plainly, this language leaves no doubt about whether the protection established by the Executive Orders applies only to entirety interests in relevant assets, or also to fractional-share interests.

The other two principal differences between the Executive Orders and the apposite Security Council resolutions are concerned with the fact that the latter apply to Iraqi oil and gas and proceeds and obligations from their sale (and the DFI), whereas the former apply as well to "any financial instruments of any nature whatsoever arising from or related to the sale or marketing thereof,"[22] and extend protection to the CBI and any of its assets, wherever located.[23] Again, with respect to the CBI, the language of section 1(a)(iii) of Executive Order 13364 extends protection to "any accounts, assets, investments, or any other property of any kind, owned by, belonging to, or held by the Central Bank of Iraq, or held, maintained, or otherwise controlled by any financial institution of any kind in the name of, on behalf of, or otherwise by the Central Bank of Iraq." The second of these two differences is self-explanatory: Because the DFI is required to hold funds from oil and gas sales in a CBI account, it makes sense to extend to CBI accounts, assets, investments, and other property the same protection accorded to the DFI. It does not matter that such are held, maintained, or otherwise controlled by another financial institution under the name or on behalf of the CBI. As for the first differences, it appears to gather within the protective umbrella of the Executive Orders all forms of creative and innovative debt and equity devices, as well as novel insurance

[20] See id.
[21] See text accompanying supra note 19.
[22] See id.
[23] See text accompanying supra note 13.

or guarantee commitments generated by or emerging from not only oil and gas sales, but also oil and gas marketing activities. It may be that the relevant Security Council resolutions reached such financial instruments by virtue to the reference to proceeds and "*obligations* arising from sales"[24] of oil and gas. With the inclusion of specific reference to protection provided for "any financial instruments of any nature whatsoever arising from or related to the sale or marketing thereof," the Executive Orders eliminate any doubt.

IV. PROTECTION APPLIED TO DIVERSE INTERESTS IN IRAQI OIL AND GAS

As discussed in the previous section, the relevant Executive Orders apply to an array of interests, instruments, and transactions that involve Iraqi oil and gas. Not only are oil and gas and the proceeds and obligations concerning them protected, the Executive Orders also protect "any financial instruments of any nature whatsoever arising from or related to the sale or marketing thereof, and interests therein."[25] Also protected is anything considered "interests therein" actual Iraqi petroleum or petroleum products. The significance of the references in the Executive Orders to "any financial instruments of any nature whatsoever" related to the sale or marketing thereof, as well as those to any "interests therein" such financial instruments, proceeds, and obligations of sales or marketing, or petroleum and petroleum products, is examined in what follows.

Addressing the latter first, recall that the matter of the entirety, as opposed to fractional-share interests, was raised from the perspective of Security Council resolutions 1483 and 1546. That same matter was again touched on in the immediately preceding section. It cannot be overemphasized, however, that the reference in the Executive Orders to "interests therein" thoroughly clarifies the fact that protection under U.S. law extends to both entirety and fractional-share interests. The effect of this is to protect from judicial proceedings any partial-share

24 Emphasis added.
25 See text accompanying supra note 19.

interest held in Iraqi petroleum, petroleum products, natural gas, proceeds and obligations, as well as any other type of financial instrument, resulting from the sale or marketing of the petroleum, petroleum products, or natural gas. When conjoined with the fact that the Executive Orders also reference "interests therein" held by any national of a foreign country other than Iraq, a situation in which a U.S. oil and gas company with a fractional-share interest in Iraqi oil and gas, or proceeds, obligations, or financial instruments arising from or related to the marketing of such, would be immune from any U.S. judicial proceedings. The fact that a U.S. company holds in the United States an instrument that represents its fractional-share, or the fact that Iraqi oil and gas in which such a company's share existed were physically present in the United States might make it otherwise subject to the U.S. judicial process. Given the immunity or protection established by the Executive Orders, action against this fractional-share interest could not proceed. Indeed, according to the precise language of the Executive Orders, were it to proceed, the result would be deemed "null and void."

Along these same lines, a situation in which an international oil and gas company contracts with the Iraqi government or one of its instrumentalities to provide an oil-field service paid for by a credit against the proceeds of the sale of oil and gas produced from the relevant field raises other interesting questions. Assume that the credit of concern was consolidated or bundled with other credits held by the company and negotiated on the international financial market. This permitted the company selling the bundled or consolidated credits to receive an immediate payment from the party to whom it was negotiated – obviously, at some reduced rate – with the purchaser of the credits being left to extract payment on the Iraqi credit at the appropriate juncture. If the purchaser of the credit held such in the United States, it might appear to be an asset of interest to one who was owed money by the purchaser. The language of the Executive Orders addressing protection for proceeds and obligations from the sale of Iraqi oil and gas, as well as "any financial instruments of any nature whatsoever" arising from or related to the sale or marketing of such oil and gas, would catch the asset represented within its protective reach. The reference in the Executive Orders to "any" financial instruments of "any nature

whatsoever" is a reference as broad and inclusive as conceivable. Regardless of the ingenuity exhibited in structuring the financial instrument concerning Iraqi oil, gas, and proceeds and obligations from their sale, the instrument is within the insulation from judicial proceedings established by Executive Orders 13303 and 13364. Neither would it matter, as observed above, that the interest targeted in a U.S. judicial proceeding was merely a fractional-share interest. The way the language of the Executive Orders is written, whether the financial instrument of concern is one in which the entire interest is held or only a partial or fractional interest, the protection against creditor action is assured. The words "interests therein" are used not only with respect to Iraqi petroleum and petroleum products, but also to proceeds, obligations, and financial instruments arising from or related to their sale or marketing.

What exactly is affected by the language of the Executive Orders that address financial instruments arising from or related to the "marketing" of Iraqi petroleum and petroleum products? Financial instruments arising from or related to the sale of such oil, gas, and products seems straightforward. As illustrated herein, the consolidation or bundling of credits – including against proceeds of Iraqi oil and gas sales – and their negotiation on the international financial markets for immediate payment at less than the face value represents a financial instrument arising from or related to the sale of oil and gas. In terms of a financial instrument arising from or related to the marketing of oil and gas, what might be envisioned is a variety of seemingly strange and yet – given the recent turmoil in international financial markets and revelations concerning the creative and highly risky instruments engineered and used therein – innovative and forward-leaning ways to generate earnings. Imagine a situation in which an instrument here styled as a "marketing default swap" were used. It might involve the creation of traded financial instruments that represent nothing more than a wager that marketing efforts regarding the disposition of Iraqi oil and gas will fetch a certain price, come in at a certain volume, or be completed or met at a certain date. In a sense, a marketing default swap is the analogue of the "credit default swap" so common in connection with the subprime real-estate meltdown. No matter the peculiar nature of the financial instrument, the broad and inclusive language of Executive Orders 13303 and 13364 makes it clear that, whether related to the sale

of Iraqi petroleum, petroleum products, or natural gas, or merely to its marketing, the instrument is insulated from the U.S. judicial process.

V. CONNECTIONS OF NATIONALITY

As discussed in Chapter 7, fractional-share interests in Iraqi oil and gas assets held by a U.S. company as compensation for technical services provided to the INOC, SOMO, or MoO are not eligible for protection from creditors as created by Executive Orders 13303 and 13364. The reason is related to the fact that the transfer of the fractional-share interest as payment for the provision of technical services constitutes a "transfer of title to an initial purchaser."[26] Conversely, protection against creditors *would* exist in a situation in which the fractional-share interest held by the U.S. company emerged from a production sharing contract/agreement (PSC/PSA) that resulted in the company collaborating with the Iraqi authorities to exploit Iraqi oil and gas – that is, the company being given a share in the oil and gas actually produced. In large measure, the fundamental distinction between the situations turns on the former scenario involving the fractional share of the oil and gas actually produced representing the compensation for services that otherwise would be paid in cash. In the latter scenario, the fractional-share interest represents an actual ownership interest in what is produced, not merely compensation for services rendered. In other words, the PSC/PSA operates on the basis of both the Iraqis and the U.S. partner company having an ownership stake in produced oil and gas long before it is ever lifted. In the case of the fractional-share interest in Iraqi oil and gas assets being payment for the provision of technical services, oil and gas owned by Iraq have been used to compensate a company with which Iraq has contracted.

When it comes to the PSC/PSA or any other situation in which there has been no transfer of title to an "initial purchaser," the Executive Orders' language that references protection from U.S. judicial process for Iraqi oil and gas assets being held by "any foreign country or a national thereof" poses no problem. The notion of "foreign country" certainly includes situations in which the Iraqi government,

[26] See text accompanying supra Chpt. 7, Sec. VI.

INOC, SOMO, MoO, or a national oil company of another nation has come into possession of Iraqi-origin oil and gas assets and holds them in the United States, perhaps for sale. It also extends not only to U.S. companies or to entities with PSC/PSA interests in Iraqi oil and gas assets, but also to situations in which those assets are held by wholly private foreign companies. For example, if the Norwegian company DNO or the Turkish company Genel Energi were to receive Iraqi PSC/PSA fractional-share interests, the terms of Executive Orders 13303 and 13364 would insulate either company from claims by creditors. Being organized under the laws of Norway and Turkey, DNO and Genel Energi are clearly considered nationals of a foreign country. The interest they hold in the Iraqi-origin oil and gas, therefore, is held by a "foreign country or a national thereof" and eligible for protection. On the other hand, if DNO or Genel Energi were only to receive Iraqi-origin oil and gas as compensation for technical services provided in oil-field development, and they then attempted to market that oil and gas in the United States, Executive Orders 13303 and 13364 would not provide protection from creditors. Although clearly considered nationals of a foreign country, the companies would be transferees of title as initial purchasers.

Again, with the addition to Executive Order 13364 section 1(a)(ii) of the language "but only until title passes to the initial purchaser," it is true that the protection from U.S. judicial process is not retained forever. That was the thrust of changing the original language of section 1(b) in Executive Order 13303 so that it provided immunity for Iraqi oil and gas and interests therein, "but only until title passes to the initial purchaser." A fractional-share interest in Iraqi-origin oil and gas held in the United States under a PSC/PSA arrangement by DNO or Genel Energi – or a U.S. company partnering with Iraq – to exploit an oil and gas field would lose immunity from judicial processes once it is sold to a purchaser. The Executive Order 13364 addition regarding the passage of title exposes transferred fractional-share interests to claims by creditors of the transferee. It is obvious that this puts transferees in precisely the same situation as a technical-service provider, whenever it receives Iraqi-origin oil and gas as compensation for services rendered.

Turning from the matter of the nationality connection evidenced by a foreign country or national thereof with an interest in an Iraqi

oil and gas asset, and examining the requirement of the Executive Orders that the particular assets of concern be "in the United States, or . . . are or hereafter come within the possession or control of United States persons," two observations are warranted. First, with regard to the assets being "in" the United States or "hereafter com[ing] within the possession" of United States persons, there can be little confusion about whether oil and gas assets are physically present in the United States. Much trickier is the second matter of whether such assets have "hereafter come within the possession" of United States persons. Before exploring that matter, commentary is offered herein on the definition of "United States person" and on the fact that the language of section 1 of the Executive Orders concerning assets "hereafter com[ing] within the possession" of U.S. persons also references "hereafter coming within the control" of such. The phrase "hereafter com[ing] within the possession" of U.S. persons stands in contrast to the relevant asset being either physically present in the United States or simply controlled by a U.S. person who is located physically outside the United States. In other words, to declare the applicability of Executive Orders 13303 and 13364 to Iraqi oil and gas assets that "hereafter come within the possession" of U.S. persons means to cover them if a U.S. person physically has the assets within his or her own grasp. This includes all such assets, even those owned by others but within the physical possession of a U.S. person. The significance of this is to insulate from any U.S. judicial process Iraq's oil and gas assets owned by the Iraqi government, INOC, SOMO, MoO, or a PSC/PSA partner and stored in tanks or holding facilities of a U.S. company – even when those tanks or holding facilities are located outside the United States. If the assets are held physically in the United States, they are within the scope of the Executive Orders by virtue of being "in" the United States. When held physically outside the United States, but in the possession of a U.S. person, they are as equally within the reach of the Executive Orders. Corroboration for this view is found in the fact that section 3(d) of Executive Order 13303 defines "Iraqi petroleum and petroleum products" to mean "petroleum, petroleum products, and natural gas originating in Iraq, including Iraqi-origin *inventories wherever located.*"[27] This emphasizes that Iraqi-origin oil and gas do

[27] See text accompanying supra note 7 at Sec. 3(d).

not lose their status as protectable assets just because they are held in inventory somewhere outside of Iraq.

Concerning the notion of a "United States person," how does one know if an Iraqi oil and gas asset is in the possession of "United States persons"? A company or enterprise organized outside the United States is not considered a "United States person," but what about a foreign subsidiary or branch of a U.S. parent company or enterprise? Do subsidiaries or branches carry the nationality of the nation-state in which they are licensed to operate; or, for the purposes of the Executive Orders, do they carry the nationality of the parent?

Section 3(c) of Executive Order 13303 defines "United States person" to mean "any United States citizen, permanent resident alien, entity organized under the laws of the United States or any jurisdiction within the United States (including foreign branches), or any person in the United States."[28] It is not inconceivable that the section 3(c) reference to "including foreign branches" could be interpreted to mean U.S.-located branches of foreign companies. However, given that section 3(c) also defines "United States person" to include "any person in the United States," then U.S. branches and subsidiaries of foreign companies would seem covered by the latter language. As a result, the phrase "including foreign branches" brings overseas branches and subsidiaries of U.S.-based parent companies within the definition of U.S. person. The result is to make Iraqi oil and gas assets held by, for example, a British- or the Netherlands-based branch or subsidiary of a U.S. company immune from U.S. judicial processes. The foreign branch or subsidiary is a U.S. person, and Iraqi assets held by U.S. persons are immune from judicial process when any foreign country or national thereof has an interest in such.

As discussed previously, the language of section 1 of the Executive Orders that references Iraqi oil and gas assets in the United States – or that hereafter come within the United States or the possession of a U.S. person – also references such assets as coming within the "control" of a U.S. person. Pointed out earlier was the distinction between "possession" and "control," the former connotes the assets being within the physical grasp of a U.S. person, and the latter not necessarily requiring

[28] See id. at Sec. 3(c).

such. Indeed, it seems that whereas the notion of "control" could encompass physical possession, it is a more inclusive concept satisfied even without actual physical possession of the relevant Iraqi oil and gas asset. A classic illustration of nonphysical possession involves the ability legally to direct the movement of Iraqi oil and gas, although it may be in the actual physical possession of another. The basic idea of Executive Orders 13303 and 13364 appears to be that until title is transferred to an initial purchaser, all Iraqi petroleum, petroleum products, and interests therein, as well as proceeds, obligations, and financial instruments arising from the sale or marketing thereof, are immune from U.S. judicial processes when the assets are in the United States or under the possession or legal control of a U.S. person. Understanding the ability to dictate the movement of a relevant asset – even though not actually exercised – as within the ambit of Executive Orders 13303 and 13364 is consistent with that idea. The widest possible range would be accorded to the protective objective of both Executive Orders. It cannot be ignored, however, that once title is transferred to an initial purchaser, immunity is lifted.

VI. A GLIMPSE AT ANALOGOUS MEASURES IN OTHER JURISDICTIONS

Having examined some of the more significant details of the U.S. domestic legal measures that affect creditor claims against Iraqi oil and gas assets, relevant provisions of both British and Canadian law provide interesting points of comparison. Although differences exist when compared side by side with Executive Orders 13303 and 13364, they are far from stark and dramatic. The relevant British measures are The Iraq (UN Sanctions) Order 2003 No. 1521 of June 14, 2003, and The Iraq (UN Sanctions) Order 2004 No. 1983 of July 29, 2004. The Canadian measure is the UN Iraq Regulations SOR/2004–221 of October 19, 2004.

British Order No. 1521 contains provisions on a variety of diverse measures. Article 9(1) of the Order provides "the Development Fund for Iraq, its property and assets, wherever located and by whomsoever held (including any rights or obligations owed by or to the Development

Fund for Iraq), shall have the like privileges and immunities as the United Nations. . . ."[29] Article 10 provides that "[a]ll petroleum, petroleum products and natural gas originating in Iraq shall have immunity from suit and legal process. Such immunity shall cease once title has passed to a purchaser."[30] With the adoption of British Order No. 1983, which was enacted subsequent to Security Council resolution 1546, article 10 of the original Order was amended by article 3 of Order 1983 to insert an article 10a in Order 1521. The inserted article 10a set forth an exemption to the rules of immunity contained not only in article 10 but also in article 9. The language of that amendment provides that "[t]he privileges and immunities provided in articles 9 and 10 shall not apply with respect to any legal proceedings, nor any resulting final judgement [sic], arising out of a contractual obligation entered into by Iraq after 30th June 2004."[31]

The language of the Canadian implementation measure is equally as fascinating. Iraq Regulations SOR/2004–221 provide in article 3(1) that "[n]o legal proceeding – and no attachment, garnishment, or execution – shall lie against (a) any petroleum, petroleum product, or natural gas originating in Iraq until the title or right to the petroleum, petroleum product, or natural gas passes to an initial purchaser; (b) proceeds or obligations from the initial sale of any petroleum, petroleum products, or natural gas originating in Iraq; or (c) the Development Fund for Iraq, including its property wherever located."[32] Article 3(2) of the Regulations follows this by providing that "[s]ubsection 1 does not apply in respect of the enforcement of a final judgement [sic] arising out of a contractual obligation entered into after June 28, 2004, by the Government of Iraq, a government of a political subdivision of Iraq, or their state bodies."[33]

Clearly, both the British and the Canadian immunity measures differ as much from one another as from the U.S. controlling Executive

[29] See The Iraq (United Nations Sanctions) Order 2003 No. 1521 (14 June 2003) at art. 9(1), available at www.opsi.gov.uk/si/si2003/uksi_20031521_en.pdf (accessed June 10, 2009).

[30] See id. at Sec. 10.

[31] See The Iraq (United Nations Sanctions) Order 2004 No. 1983 (29 July 2004) at art. 3, available at www.opsi.gov.uk/si/si2004/uksi_20041983_en.pdf (accessed June 10, 2009).

[32] See United Nations Iraq Regulations SOR/2004–221 (Oct. 19, 2004) at art. 3(1), available at http://laws.justice.gc.ca/PDF/Regulation/S/SOR-2004–221.pdf (accessed June 10, 2009).

[33] See id. at art. 3(2).

Orders. Most conspicuous, article 3 of British Order 1983, which amends the principal Order to add an article 10a, speaks in terms of a lifting of privileges and immunities with respect to contractual obligations of Iraq incurred after June 30, 2004, while article 3(1) of the Canadian Iraq Regulations SOR/2004–221 references June 28, 2004, as the instrumental date for loss of immunity. Additionally, those same provisions differ on the matter of immunity from legal action, as opposed to UN-like privileges and immunities. The principal British Order establishes immunity for petroleum, petroleum products, and natural gas (until title is passed), and accords the DFI and its property and assets the same privileges and immunities as the UN. The Canadian Iraq Regulations simply indicate immunity from legal proceedings for Iraqi petroleum, petroleum products, and natural gas (until title is passed), proceeds and obligations arising from their sale, and the DFI and its property. No indication appears in the Canadian Regulations that the DFI is entitled to UN-like privileges and immunities. Furthermore, as just suggested, article 3(1) of the Canadian Regulations provides protection for proceeds and obligations from the sale of oil and gas, whereas no such language appears in the provisions of the British Orders. Article 3(1)(b) of the Canadian Regulations explicitly refers to "proceeds and obligations from the initial sale of any petroleum, petroleum products, or natural gas originating in Iraq"[34] as protected from legal action. Such language is absent from British Orders 1521 and 1983.

As already seen, the U.S. Executive Orders clearly extend protection from legal action to proceeds and obligations from the sale of Iraqi oil and gas. To this extent, they align more with the Canadian Regulations than with the British Orders. Recall, however, that the protection accorded by the U.S. Executive Orders is protection from "judicial process." The Canadian measure, on the other hand, establishes protection from "legal proceeding[s]"; the British measures protect petroleum, petroleum products, and natural gas from "legal process," and accord the DFI the same "privileges and immunities" as the UN. This difference suggests more breadth to the direct thrust of both the Canadian and British measures. But given the fact that even decisions and determinations deriving from arbitral or administrative

[34] See id. at art. 3(1)(b).

(i.e., nonjudicial) proceedings typically result in subsequent enforcement action in the judiciary, the significance of any distinction between the direct thrust of the Canadian and British measures and the U.S. Executive Orders may not be especially meaningful.

Without getting into a lengthy examination of the matter, a major difference between the Canadian and British measures and U.S. Executive Orders 13303 and 13364 involves the entire question of a nationality connection. As observed previously, the U.S. Executive Orders envision such, both with respect to the insulated or protected Iraqi oil and gas, interests therein, and proceeds and obligations arising from their sales having to be "in the United States" or "within the possession or control of United States persons," as well as with respect those very Iraqi oil and gas assets being assets "in which any foreign country or a national thereof has any interest."[35] From the way that the Canadian Iraq Regulations SOR/2004–221 and British Orders 1521 and 1983 are written, a nationality connection of either sort is not explicitly required. It is not likely that courts or other dispute-settlement bodies in Canada or Great Britain would be excited about exercising jurisdiction in instances in which the ultimate object sought by way of satisfaction for a creditor claim is physically situated outside the borders of the country and beyond the possession or control of an entity over which adjudicative authority exists. Consequently, even without explicit language that requires a nationality connection, it may arise implicitly. Regarding the Iraqi oil and gas assets being assets in which a foreign country or national thereof has an interest, it is difficult to know precisely how Canadian and British courts or dispute-settlement bodies might react. The language on that matter in the U.S. Executive Orders expands their insulation or protection to cover such assets, even though Iraq or its governmental organs may have no interest therein. In the absence of that type of clarification, it is uncertain whether the Canadian and British measures will be interpreted to apply only when the Iraqi government or one of its nationals has an interest in the relevant oil and gas assets.

It also bears noting that the Canadian and British measures do not provide for the coverage of financial instruments arising from or related to the sale or marketing of Iraqi oil and gas or interests in them. The

[35] See supra Sec. V.

U.S. Executive Orders, however, make explicit that such are clearly within the protective effect of Executive Orders 13303 and 13364.[36] This departs from the plain language of the relevant Security Council resolutions, which reference only Iraqi oil and gas, and proceeds and obligations arising from their sale. Tracking the specific language of the UN resolutions, the Canadian Iraq Regulations SOR/2004–221 refer to Iraqi petroleum, petroleum products, or natural gas, and proceeds or obligations from their initial sale.[37] The language of British Order 1521 – the principal Order on the matter of immunity – contains a slightly more interesting formulation, given that any reference to proceeds and obligations arising from the sale of oil and gas is not only cast in different words, but also is connected to the DFI. Recall that both Security Council resolutions speak of petroleum, petroleum products, and natural gas as being immune from proceedings, and proceeds and obligations arising from sales thereof (and the DFI) as being entitled to UN-like privileges and immunities. The U.S. Executive Orders reference petroleum, petroleum products, natural gas, and interests therein as not being subject to judicial processes, and proceeds, obligations, and financial instruments associated with their sale or marketing (and the DFI) as being in the same position. The language of British Order 1521, by way of contrast, indicates in article 10 that petroleum, petroleum products, and natural gas have immunity from suit and legal process; and, in article 9, that the DFI and property and assets "including any rights or obligations owed by or to the DFI"[38] have the same privileges and immunities as the UN. It is obvious that each domestic-law formulation opts for a unique approach to implementing the controlling resolutions of the Security Council.

VII. CONCLUSION

The intricacies and complexities of the domestic-law measures in the United States and other nations that are designed to implement Security Council resolutions 1483 and 1546 – as extended by resolution 1905 of December 2009, and before that, resolution 1859 of December

[36] See supra Sec. IV.
[37] See United Nations Iraq Regulations, supra note 32 at art. 3(1)(a-b).
[38] See The Iraq (United Nations Sanctions) Order 2004 No. 1521, supra note 29.

2008 – are many and extremely involved. In most respects, U.S. Executive Orders 13303 and 13364 parallel the relevant Security Council resolutions in extending protection from creditor claims to Iraqi oil and gas assets and those charged with managing and administering them. In various different respects, however, the Executive Orders contain language that can be interpreted as substantially enhancing the protection called for by the UN. If the preceding statements constitute an accurate assessment of the factual situation, questions may be raised regarding foundational support for the issuance of such having to derive from something other than the adoptions of the Security Council. These questions would necessitate a close look at the authority of the President provided for in U.S. legal enactments (e.g., the International Emergency Economic Powers Act), which are outside the scope of this study and therefore not examined here.

Aside from issues associated with the specifics of the U.S. Executive Orders concerning the protection and insulation provided to Iraqi oil and gas assets, a comparison with domestic measures of selected other countries suggests differing approaches to implementing the Security Council resolutions. In some instances, the differences may be semantic and merely cosmetic but not of any real practical significance. In other instances, the differences may provide the opportunity for meaningful divergence from what is done elsewhere to protect Iraqi oil and gas assets. It is clear that the review in this chapter highlights the most apparent and interesting issues related to U.S. and other domestic-law measures protecting Iraqi assets from creditors desirous of securing satisfaction for what they consider to be owed them. Nevertheless, if the language of the relevant domestic measures in each jurisdiction were sedulously winnowed, the results would fill multiple volumes. In exploring just below the surface, thereby permitting the nuances and assorted angles of the various domestic measures to emerge, the objective has been simply to increase the legal community's attention to the details of the national legal adoptions aimed at giving effect to UN Security Council resolutions 1483 and 1546.

9 RECOGNITION AND ENFORCEMENT OF DETERMINATIVE ACTIONS AFFECTING IRAQI OIL AND GAS ASSETS

I. INTRODUCTION

As discussed in Chapter 8, there may be language in the provisions of national measures designed to insulate Iraqi oil and gas assets from claims by creditors that could be susceptible to the interpretation of providing protection only from court action and not from administrative, arbitral, or other nonjudicial action. To illustrate this point, recall that U.S. Executive Orders 13303 and 13364, as confirmed and extended by the presidential Notice of May 20, 2009, provide only for insulation from "judicial process." As also indicated, however, even in situations in which the language of the relevant national measures is limited to court proceedings, determinations or decisions emerging from administrative, arbitral, or other nonjudicial proceedings are typically not self-enforcing and therefore require subsequent court action seeking recognition and enforcement to give them any real impact. A clear distinction exists between national measures insulating Iraqi oil and gas assets from any proceeding or process aimed at determining legal rights, and those written in such a way that they protect only against court proceedings. However, to the extent that the latter measure is involved, Iraqi assets are exposed to substantial inconvenience and problems by possibly having to be defended against nonjudicial proceedings. Ultimately, given the need for determinations from administrative or arbitral proceedings to be buttressed by recognition and enforcement in the judicial arena, the difference between both sorts of measures is minimal.

The foregoing may be based on the possibility of a situation in which a particular national measure is cast in language that permits

maintaining nonjudicial proceedings against targeted Iraqi oil and gas assets. However, with subsequent court recognition and enforcement being sought in the same nation-state, the possibility of a determination or decision from proceedings initiated and completed in one nation being brought to another for recognition and enforcement cannot be dismissed. This could result from at least one of two situations, which might involve (1) the judgment or determination of a nonjudicial proceeding from one country brought to the courts of another for recognition and enforcement; or (2) the courts of one national jurisdiction being asked to recognize and enforce an actual court decision from another.

The second scenario may seem anomalous in view of the controlling Security Council resolutions directive that all member states implement the protective goals announced by the Council. Nonetheless, as has been seen, the goals are couched in language that is subject to a certain amount of interpretation, and it is possible that the implementing measures adopted by some member states could provide for a degree of latitude concerning court action not available in others. Furthermore, in light of the certainty that all nations move legislatively with varying degrees of celerity, and that the constitutional peculiarities of some nations preclude foreclosing access to the courts for prosecuting creditor claims against the assets of any nation, some UN member states may have yet to adopt measures precluding action against Iraqi oil and gas assets, and others may feel constitutionally compelled to refrain from doing so. Additionally, it cannot be ignored that from among the UN member states, at least some may have consciously decided to forgo implementing the directive to insulate Iraqi oil and gas assets. This study has not surveyed the laws of each member state to determine whether particular nations make that decision because of an overwhelming sense of legal entitlement, strong antipathy toward the existing Iraqi governing regime, or feelings of the absence of fair dealing. However, it is conceivable that deliberate refusals to execute the mandates of resolutions 1483 and 1546 number among UN member states.

With respect the former situation – that involving a determination or decision in a nonjudicial proceeding from one nation being brought to another's courts for recognition and enforcement – this is

undoubtedly the far more likely to occur and, therefore, is the only one focused on in Section IV. After all, most nations must be considered as taking seriously their Security Council resolution obligations; they are not likely to scoff at implementing directives articulated in those resolutions. It is possible that other legislative- and executive-branch work could slow the speed at which implementing measures ultimately are adopted, and it is also clear that the terms of resolutions 1483 and 1546 have been construed in a variety of slightly different ways. Nevertheless, most nations would seem to feel a need to avoid being perceived as in complete noncompliance with Security Council mandates. Given the unequivocal intent expressed in the resolutions about the insulation of Iraqi oil and gas assets from being jeopardized by legal action, it thus seems unlikely that more than a few instances of courts in one nation-state being asked to recognize and enforce court judgments, determinations, or decisions of another would occur. The more likely scenario involves requests for recognition and enforcement of nonjudicial determinations and decisions.

The second point that needs to be made by way of introduction concerns the strict language of paragraph 22 of Security Council resolution 1483 and the various national implementing measures. That language relates to the distinction between the protection provided to Iraqi petroleum, petroleum products, and natural gas, as opposed to the protection provided to proceeds and obligations arising from their sale. The fact that paragraph 22 of resolution 1483 provides immunity from legal proceedings and insulation from attachment, garnishment, and execution for Iraqi oil and gas, whereas proceeds and obligations arising from their sale are extended UN-like privileges and immunities, has been discussed in detail in preceding chapters. To illustrate the national implementing measures, section 1 of Executive Order 13303 speaks instead of protection from attachment, judgment, decree, lien, execution, or other judicial process and makes it applicable to Iraqi petroleum, petroleum products, and natural gas, and interests therein, as well as proceeds and obligations and other financial instruments arising from or related to their sale or marketing. Section II of this chapter integrates this distinction and focuses on nonjudicial determinations or decisions concerning Iraqi oil and gas and presented for recognition and enforcement before the courts of the rendering nation-state.

Section III thereafter takes up the matter of recognition and enforcement in the same jurisdiction of nonjudicial determinations or decisions involving proceeds and obligations and other interests associated with the disposition of Iraqi oil and gas. Section IV focuses on issues generated by requests for recognition and enforcement of nonjudicial determinations or decisions when placed before the court of a foreign state – a state other than the one that rendered the judgment for which the request was submitted. Because the possibility that court judgments from one jurisdiction will be placed before the courts of another nation and recognition and enforcement requested is minimal, the matter is discussed only briefly in Section IV, and exclusively in the context of requests for foreign courts to recognize and enforce nonjudicial determinations and decisions of another state.

II. DETERMINATIONS OR DECISIONS IN NONJUDICIAL PROCEEDINGS INVOLVING IRAQI OIL AND GAS AND PLACED BEFORE COURTS IN THE NATION OF THE RENDERING JURISDICTION

Assume that the national implementing measure of a nation-state committed to honoring its international obligation concerning Security Council resolutions 1483 and 1546 provides, as in the United States, for Iraqi oil and gas immunity from "judicial process," and that the provision is understood as not protecting such from actions initiated before nonjudicial entities. In such a situation, is the matter of a determination or decision resulting from the nonjudicial process involving Iraqi oil and gas and placed before a court of the same country one that is addressed by the language of the relevant international and domestic measures? To restate the query, what is offered by the precise terms of the controlling Security Council resolutions and national implementing measures that addresses whether nonjudicial determinations or decisions involving Iraqi oil and gas and issued within one country are to be recognized and enforced by courts within that same country? As suggested in Chapter 8, even if a national measure concerning the protection of Iraqi oil and gas permitted actions to proceed before nonjudicial bodies, the fact that such a measure precludes action before judicial bodies offers substantial protection by eliminating the ability

of any creditor to secure a favorable judgment in an administrative or arbitral proceeding – for example, converting that judgment into full and effective satisfaction before a court of law.

Regarding the language of the controlling Security Council resolutions, recall that paragraph 22 of resolution 1483 (as amended and extended by resolutions 1546, 1859, and 1905) references Iraqi petroleum, petroleum-product, and natural-gas insulation from "legal proceeding[s]" and "any form of attachment, garnishment, or execution."[1] It will also be recalled that both the May 9 and May 15 drafts of what became paragraph 22 spoke of petroleum, petroleum-product, and natural-gas insulation from "judicial, administrative, arbitration, or any other proceedings (including any prejudgment or post-judgment attachment, garnishment, execution, or any other action to satisfy judgment)."[2] Before Section III offers analysis, recall that paragraph 22 speaks of proceeds and obligations connected with the sale of Iraqi oil and gas being entitled to the same privileges and immunities that are extended to the UN. Again, this makes clear the distinction between the protections accorded Iraqi petroleum, petroleum products, and natural gas, on the one hand, and proceeds and obligations arising from their sale, on the other.

As discussed previously, the language of the draft formulations are clear in precluding any sort of legal action against Iraqi oil and gas proper, whether before a court, an administrative body, or an arbitral panel.[3] Nonetheless, the linguistic configuration of immunity from "legal proceedings" selected in paragraph 22 of resolution 1483 can be accorded the same kind of reading. Of greatest relevance for the matter of potential court enforcement of any nonjudicial determination or decision rendered in the same nation-state and involving Iraqi oil and gas, however, is the fact that both the May 9 and May 15 drafts spoke in terms of immunity from not only prejudgment or postjudgment attachment, garnishment, or "execution," but also "any other action to satisfy judgment." The implication, of course, is that not only are usual court processes of attachment and garnishment foreclosed, but so, too, are processes that aim at "execution" of judgments or involve other "action to satisfy" such. Paragraph 22 of resolution

[1] See text accompanying supra Chpt. 3, note 21.
[2] See text accompanying supra Chpt. 3, notes 33–34.
[3] See text accompanying supra Chpt. 7, Sec. VI.

1483 ignores the reference to "any other action to satisfy judgment," but it retains the language referencing "execution" of judgments. To the extent that the concept of "execution" simply evidences Iraqi oil and gas immunity from actions that seek satisfaction for determinations or decisions targeting those assets, there is good reason to believe that it is intended to foreclose the possibility of subsequent court action to recognize and enforce administrative or arbitral opinions targeting those assets. There can be little dispute that the language of the earlier draft resolutions may have articulated more explicitly that position. What was settled on by the Security Council delegates in the penultimate version of paragraph 22, though, appears more than adequate to accomplish the same objective. After all, the notion of a prohibition on "execution" – when expressed in the context of a general prohibition on attachment, garnishment, and legal proceedings – indicates foreclosure of any action designed to secure satisfaction by a claimant against the assets of a defendant. Corroborative of this interpretation is the fact that the language of paragraph 22 couches its preclusion of attachment, garnishment, or execution with the words "any form" of attachment, garnishment, or execution. The breadth of the reference to "any form" certainly connotes immunity from attachment, garnishment, or execution, regardless of whether it takes the form of a direct court execution, or an indirect execution by virtue of court recognition and enforcement of a noncourt, nonjudicial determination or decision.

With respect to what is said in the U.S. Executive Orders 13303 and 13364, they are explicit in precluding "judicial" process alone. This language does not speak directly to the recognition and enforcement of determinations or decisions rendered initially in nonjudicial proceedings. Indirectly and inferentially, the foreclosing of access to "judicial" process would seem to eliminate an administrative or arbitral judgment subsequently being recognized and enforced in a court of law. Other language in section 1 of Executive Order 13303 confirms this interpretation by specifically indicating Iraqi petroleum, petroleum-product, and natural-gas immunity from "attachment, judgment, decree, lien, execution, garnishment" or "other judicial process." (It also indicates such immunity for proceeds and obligations, and other financial interests arising from or related to the sale or marketing of Iraqi petroleum, petroleum products, or natural gas; a matter taken up below in Section III.) The language of section 1 in Executive Order 13303 is

plainly an effort to be more explicit than the language of paragraph 22 in Security Council resolution 1483. Rather than referencing Iraqi oil and gas immunity only from attachment, garnishment, or execution, section 1 references immunity from "judgment," "decree," and "lien" as well. When coupled with the further reference to "or other judicial process," requests for courts to recognize and enforce nonjudicial determinations and decisions rendered in the same jurisdiction appear precluded. In addition to the notion of no judicial "execution" – a notion common to both Security Council resolution 1483 and Executive Order 13303 and protecting against recognition and enforcement actions – section 1 of Executive Order 13303 protects Iraqi oil and gas proper against "judgment[s]" or "decree[s]" and any "other judicial process." Neither the term *judgment* nor *decree* is defined by the Executive Order; however, a prohibition on "judgment[s]" or "decree[s]" against Iraqi oil and gas would have the effect of preventing U.S. courts from converting nonjudicial determinations or decisions into binding and consequential action exhibiting meaningful and strong results. The section 1 references to "judgment" and "decree" can be understood as foreclosing the full spectrum of court action, including that simply recognizing and enforcing actions of nonjudicial entities. Supporting such an understanding of these terms is the additional reference in section 1 to the fact that not only are judgments and decrees, as well as attachments, liens, and garnishments prohibited, so also are "other judicial process[es]." Actions seeking recognition and enforcement of things like administrative or arbitral decisions cannot avoid being perceived as judicial processes other than attachment, judgment, decree, lien, execution, or garnishment.

Prior to commenting on both the British and Canadian domestic measures and their implications for recognition and enforcement of nonjudicial actions against Iraqi oil and gas itself, a few points require reiteration. First, the situations discussed in Section II address only recognition and enforcement in a single jurisdiction – the courts of the same state rendering a nonjudicial determination or decision are asked to recognize and enforce it. The complexities of foreign-state recognition and enforcement are taken up in Section IV. Second, as is apparent from the discussion herein, the language of both paragraph 22 of Security Council resolution 1483 and section 1 of Executive Order 13303 examined in Chapter 8 in connection with the matter of the nature of

the proceedings foreclosed against Iraqi oil and gas assets provides the focus of this section. However, the concepts of "legal proceeding" in resolution 1483 and "judicial process" in Executive Order 13303 previously proved instrumental. Presently, it is the accompanying concepts of "attachment, garnishment, or execution" and "attachment, judgment, decree, lien, execution, and garnishment" that are central. The third point concerns the relationship between what had earlier been said about the language of measures that protect only against "judicial" process or proceedings, and the significance of thereby permitting nonjudicial actions to move forward. In particular, it was observed that permitting the actions to move forward has little ultimate consequence. Yes, there is inconvenience with the potential of having to mount a defense in an administrative or arbitral proceeding. Ultimately, however, the resulting judgments remain ineffective in accessing Iraqi oil and gas assets, as long as judicial action is foreclosed. The present section focuses on the associated matter of whether the language of the relevant UN or national implementing measures addresses possible recognition and enforcement. As observed, both resolution 1483 and Executive Order 13303 appear to preclude such, although not in unequivocal and straightforward language. Fourth, to reiterate, Section II focuses on the protection accorded only Iraqi petroleum, petroleum products, and natural gas; not until later is anything of consequence offered about the protection accorded the proceeds and obligations arising from their sale.

As for the British and Canadian measures and the protection provided to Iraqi oil and gas in regard to recognition and enforcement actions, recall that the Canadian measure SOR/2004–221 contains relevant language that tracks what is stated in Security Council resolution 1483. Specifically, the relevant language states that Canada permits no "legal proceeding" or "attachment, garnishment, or execution" against Iraqi oil and gas assets.[4] It is clear that a difference exists between foreclosing "judicial" process or proceedings, and foreclosing "legal" process or proceedings. To the extent that the latter is capable of encompassing both court and noncourt actions, its use seems significant. Of even greater importance in terms of recognition and enforcement, however, is the fact that the Canadian measure also employs the term

[4] See text accompanying supra Chpt. 8, note 32.

execution in the phrasing of its protection of Iraqi oil and gas. This is in the reference in article 3(1) to the fact that "[n]o legal proceeding – and no attachment, garnishment, or execution – shall lie against (a) any petroleum, petroleum product, or natural gas originating in Iraq until the title or right to the petroleum, petroleum product, or natural gas passes to an initial purchaser." If the term *execution* is sufficient under Security Council resolution 1483 (and Executive Order 13303) to eliminate the potential to seek judicial recognition and enforcement of nonjudicial determinations or decisions rendered in the same jurisdiction, then it is sufficient under section 3(1) of Canadian measure SOR/2004–221.

British Order 1521, the operative measure regarding insulation of Iraqi oil and gas assets, uses entirely distinct language by referencing immunity from "suit" and "legal process"[5]; there is no indication about immunity from attachment, garnishment, execution, judgments, liens, or decrees. Despite the latter absence, however, to preclude access to "suit" or "legal process" is sufficient to foreclose action to recognize and enforce nonjudicial determinations and decisions against Iraqi oil and gas proper. Declaring that there will be no "suit" against or "legal process" instituted regarding Iraqi oil and gas seems just as effective at stopping any suit or legal process brought to secure court recognition and enforcement of an earlier nonjudicial determination or decision, as it is at stopping a direct, original action against such. Thus, whereas both the Canadian and British measures share with Security Council resolution 1483 and Executive Order 13303 an absence of explicit language clearly precluding actions aimed at securing recognition and enforcement of administrative or arbitral determinations or decisions, the language employed appears more than up to the task.

III. RECOGNITION AND ENFORCEMENT OF NONJUDICIAL DETERMINATIONS OR DECISIONS INVOLVING PROCEEDS AND OBLIGATIONS FROM OIL AND GAS DISPOSITIONS

As previously discussed herein, the language of paragraph 22 of Security Council resolution 1483 distinguishes between the insulation or

[5] See text accompanying supra Chpt. 8, note 30.

protection provided Iraqi oil and gas, on the one hand, and the proceeds and obligations that can arise from their sale, on the other. Iraqi petroleum, petroleum products, and natural gas are immune from legal proceedings, and from attachment, garnishment, or execution. Proceeds and obligations associated with the sale of the oil and gas are under the same privileges and immunities as the UN (as is the DFI). The essential questions explored in this section are thus: whether any significance attaches to this distinction; and, whether the same type of distinction (with accompanying significance) exists relative to the national implementing measures of the United States, Canada, and Great Britain. There is little doubt that according UN-like privileges and immunities to the proceeds and obligations arising from Iraqi oil and gas sales (and to the DFI) insulates them from claims brought by creditors – at least to the same extent as the insulation received by the oil and gas.[6] However, does it make a difference when it comes to recognition and enforcement of nonjudicial determinations and decisions that oil and gas is spoken of in terms of immunity from legal proceedings and attachment, garnishment, and execution? Additionally, given that the language employed in the various national implementing measures differs – and, in some instances, even from the controlling Security Council resolution – is it possible that, in the context of recognition and enforcement requests, the protection extended to proceeds and obligations associated with Iraqi oil and gas disposition varies depending on the national measure concerned?

To review the bidding, as seen before, paragraph 22 of resolution 1483 refers to petroleum, petroleum products, and natural gas as being "immune" from legal proceedings, and any form of "attachment, garnishment, or execution," with proceeds and obligations arising from their sale (and the DFI) enjoying "privileges and immunities" equivalent to those of the UN. In contradistinction, section 1 of Executive Order 13303 (amended by Order 13364 and extended by presidential Notice of May 20, 2009) declares simply that "any attachment, judgment, decree, lien, execution, garnishment, or other judicial process" is prohibited and null and void against Iraqi "petroleum, petroleum products . . . and proceeds, obligations, or any financial instruments . . . arising from or related to the sale or marketing thereof . . . "

[6] See text accompanying supra Chpt. 7, Sec. III.

(and the DFI and CBI or its assets or accounts). Article 3(1) of Canadian Iraq Regulations SOR/2004–221 provides "[n]o legal proceeding – and no attachment, garnishment, or execution – shall lie against (a) any petroleum, petroleum product, or natural gas originating in Iraq . . . ; (b) proceeds or obligations from the initial sale of any petroleum, petroleum product, or natural gas originating in Iraq; or (c) the Development Fund for Iraq, including its property wherever located." British Order 1521 provides in article 10 that Iraqi petroleum, petroleum products, and natural gas have "immunity from suit and legal process" and article 9(1) indicates that the DFI, "its property and assets wherever located and by whomsoever held (including any rights or obligations owed by or to the Development Fund for Iraq), shall have the like privileges and immunities as the United Nations. . . ." As is well known, the protective effect of these measures is lifted with respect to judgments arising from Iraqi-government contractual obligations entered into after the CPA handover of power in June 2004. Consistent with that, British Order 1983 amended article 10 to include article 10a that declares "[t]he privileges and immunities provided in articles 9 and 10" shall not apply with respect to such judgments.

Before examining the importance of these provisions in connection with recognition and enforcement of nonjudicial determinations or decisions related to proceeds and obligations, two observations are required. One deals with the fact that British Order 1983, in adding an article 10a to the original language of article 10 of Order 1521, characterizes both articles 9 and 10 as providing for privileges and immunities when, in point of fact, article 9(1) grants privileges and immunities to the DFI and its assets, but article 10 references only that "immunity from suit and legal process" be extended to Iraqi oil and gas. The other observation concerns the Canadian measure; specifically, article 3(1) includes nothing at all about privileges and immunities – the only reference is to immunity in general. The article 3(1) chapeau begins by stating that legal proceedings, as well as attachment, garnishment, or execution, are prohibited. It then continues in three subsequent subparagraphs to identify exactly what is protected by that prohibition, naming Iraqi oil and gas, proceeds and obligations from the sale of such, and the DFI.

With these two observations in the back of one's mind, it is clear that paragraph 22 of Security Council resolution 1483 distinguishes

between immunity and UN-like privileges and immunities. Concerning whether this distinction proves significant in the context of requests for recognition and enforcement of nonjudicial determinations and decisions regarding proceeds and obligations associated with dispositions of Iraqi oil and gas, recall that paragraph 21 of the May 9 draft of what became paragraph 22 of Security Council resolution 1483 provided that Iraqi "petroleum, petroleum products, and natural gas . . . and proceeds of sales thereof shall be immune from judicial, administrative, arbitration or any other proceedings (including prejudgment or post-judgment attachment, garnishment, or execution or other action to satisfy judgment)," with the UN-like privileges and immunities only being mentioned in the paragraph 15 reference to the so-called Iraqi Assistance Fund, proposed predecessor of the DFI.[7] Paragraph 19 in the follow-on May 15 draft had the same provision, with one inconsequential difference concerning immunity for "indebtedness incurred in connection with . . . [oil and gas] sales."[8] In other words, both drafts distinguished between general immunity from creditor action and UN-like privileges and immunities, but neither applied that distinction in a way that would have accorded Iraqi oil and gas immunity, and the proceeds and obligations arising from their sale privileges and immunities equivalent to those of the UN.

With respect to the significance of the fact that what ultimately became the language of paragraph 22 of resolution 1483 applies just such a distinction with regard to oil and gas and the proceeds and obligations arising from their sale, the significance apparently is inconsequential when it comes to requests for court recognition and enforcement of nonjudicial determinations or decisions rendered in the jurisdiction of the petitioned judicial body. As suggested in Chapter 7, it is one thing to accord Iraqi oil and gas itself immunity from legal proceedings, and something considerably different to extend proceeds and obligations from their sale (and the DFI) UN-like privileges and immunities. The latter notion reaches well beyond protection from action designed to determine legal rights, and includes protection from not only judicial-branch, but also executive- and legislative-branch and other governmental action. This seems clear from the fact of the 1946

[7] See text accompanying supra Chpt. 7, note 7, and Chpt. 3, note 33.
[8] See text accompanying supra Chpt. 7, note 8, and Chpt. 3, note 34.

Convention on the Privileges and Immunities of the United Nations, which was adopted shortly after the UN Charter and was targeted at giving effect to the Charter's language in article 105 that guaranteed full privileges and immunities to that international body.[9] From this, it seems logical to infer that, because Iraqi-origin oil and gas are accorded immunity from legal proceedings, and proceeds and obligations from their sale are accorded the more extensive privileges and immunities equivalent to those enjoyed by the UN, since nonjudicial determinations and decisions involving oil and gas proper are not to receive recognition and enforcement in courts the rendering jurisdiction, then the same result should obtain with regard to requests for recognition and enforcement of those determinations or decisions involving proceeds and obligations. If administrative or arbitral judgments involving Iraqi petroleum, petroleum products, or natural gas are not eligible to be recognized and enforced, then why should the more protected proceeds and obligations resulting from their sale be treated differently?

That the national implementing measures of the United States, Canada, and Great Britain, for example, may be seen as departing in certain respects from the directive of paragraph 22 of resolution 1483 is not surprising. The unclear language of the resolution leaves much interpretive latitude, and UN member states will always bring divergent policy perspectives to their individual efforts to fulfill their implementation responsibilities. Can anything be achieved from an analysis of the relevant language of U.S. Executive Order 13303, Canadian Iraq Regulations SOR/2004–221, and British Order 1521, in terms of insights they offer on the question of recognition and enforcement of nonjudicial determinations or decisions involving proceeds and obligations associated with Iraqi oil and gas dispositions?

As pointed out very near the beginning of this section, the language of section 1 of Executive Order 13303 makes no reference at all to the notion of UN-like privileges and immunities, preferring instead to subject Iraqi oil and gas, as well as proceeds and obligations and financial instruments associated with their sale or marketing (and the DFI), to general immunity from attachment, judgment, decree, lien, execution, garnishment, or other judicial process. The absence of the extension of privileges and immunities of the UN to Iraqi oil and gas sales proceeds

[9] See text accompanying supra Chpt. 7, notes 14–16.

and obligations may fail to fully effectuate the protection mandated by paragraph 22 of Security Council resolution 1483. Nonetheless, concerning the present issue of recognition and enforcement relative to proceeds and obligations, it is clear that there is no better chance of securing such with respect to proceeds and obligations, than with respect to Iraqi oil and gas proper. The language of section 1 of Executive Order 13303 is explicit in undercutting the chance for recognition and enforcement of a nonjudicial determination or decision against either type of Iraqi asset. Not only are "judgment[s]" and "execution[s]" prohibited, but so, too, are all "other judicial process[es]." In prohibiting "execution[s]," the Executive Order attempts to tap into the idea that even actions seeking to secure some compensatory or punitive result or impact on assets, properties, or interests held by one who has suffered an adverse judgment are precluded. By their very nature, then, this forecloses the possibility of any recognition and enforcement of any type of determination or decision.

Recall the observations about the differences between the Canadian and British national implementing measures. Article 3(1) of the Canadian measure – similar to the U.S. Executive Order – makes no distinction between oil and gas proper, and proceeds and obligations from their sale, and places both under general immunity from legal proceedings and from attachment, garnishment, or execution. British Order 1521 employs the notion of privileges and immunities, as well as general immunity, but applies the former to the DFI and its property, assets, and accounts, and the latter to Iraqi oil and gas – and then inserts article 10a in British Order 1983 that confuses the matter of whether a true distinction exists between the two notions.

On the Canadian measure, it provides that no "legal proceeding" shall lie against Iraqi petroleum, petroleum products, and natural gas until title has passed to a purchaser, or against the proceeds or obligations of initial sales thereof (or the DFI). As suggested, there is ample reason to believe that that language can be interpreted as even foreclosing the possibility of nonjudicial proceedings to settle legal rights. Nevertheless, if it is read strictly as meaning only to prevent court action, thereby allowing nonjudicial action to proceed, any judgment in such nonjudicial action eventually may be submitted to a court with a request for recognition and enforcement. While offering nothing in particular regarding recognition and enforcement of nonjudicial

determinations or decisions when it comes to proceeds and obligations, the prohibition on "legal proceedings" set forth in article 3(1) of Canada's UN Iraq Regulations SOR/2004–221 seems adequate to prevent them. By insulating Iraqi oil and gas as well as proceeds and obligations from their sale from any "legal proceeding," the language of the article prevents someone with a determination or decision in an administrative, arbitral, or other nonjudicial action from seeking recognition and enforcement before a court of law.

The plain language of the provision states that "[n]o legal proceeding . . . shall lie" and, in doing so, encompasses all legal proceedings, whether instituted in the first instance before a court, or brought before a court through a request for subsequent recognition and enforcement of a noncourt determination or decision. Furthermore, it is inaccurate to suggest that the "legal proceeding[s]" prohibited by article 3(1) are confined to those involving "attachment, garnishment, or execution," all of which also are explicitly precluded by the terms of article 3(1). The reason for this is that "attachment, garnishment, and execution" are mentioned as additional actions prohibited against Iraqi oil and gas and the proceeds and obligations from their sale (and the DFI), not as exhaustive of the legal proceedings foreclosed. Again, the language states: "No legal proceeding – and no attachment, garnishment, or execution – shall lie against" oil and gas of Iraqi origin or proceeds or obligations associated with their sale. Even if the nature of the limited legal proceedings was defined as those involving attachment, garnishment, or execution, in light of the fact that recognition and enforcement actions – including those related to nonjudicial determinations or decisions concerning proceeds and obligations – necessarily involve "execution" of an antecedent judgment in order to access assets, property, or other interests of a defendant, the language of article 3(1) clearly establishes a prohibition.

Recognition and enforcement of nonjudicial judgments targeting proceeds and obligations of the sale of Iraqi oil and gas present interesting questions under the relevant British national implementing measures. Regarding the question involving the British Order 1983 addition of an article 10a to the earlier article 10 of Order 1521, it speaks of "[t]he privileges and immunities provided in articles 9 and 10" as not applying with respect to post-CPA handover contractual obligations. As discussed previously, however, the only reference to "privileges

and immunities" in the earlier British Order 1521 is in the article 9 characterization of protection to be accorded the DFI and its assets, property, and accounts. They are to be subject to the same "privileges and immunities" enjoyed by the UN. Article 10, on the other hand, references only "immunity" from suit and legal process in the statement of protection accorded Iraqi oil and gas; there is nothing in that provision about "privileges and immunities."

As indicated earlier, the notions of "immunity" from legal process or proceedings, and "privileges and immunities" equivalent to those of the UN, are far from coterminous or of equal import. Although the latter notion offers greater protection, it is just as effective in foreclosing the possibility of legal action as the mere notion of immunity. It is certainly possible that the addition of article 10a effected by British Order 1983 conflates the "immunity" from suit and legal process accorded Iraqi oil and gas by the earlier article 10 of Order 1521 so that it is the equivalent of the "privileges and immunities" granted to the DFI and its assets, property, and accounts. It is equally possible that the new article 10a reference to the "privileges and immunities provided in articles 9 and 10" simply indicates that the privileges and immunities of the DFI, and the mere immunities of Iraqi oil and gas, do not preclude legal proceedings arising from contractual obligations of the Iraqi government entered into after the CPA's transfer of power to indigenous Iraqi elements.

Whatever is the proper interpretation of the British Orders' language regarding that matter, it is clear that article 10 of Order 1521 accords Iraqi oil and gas "immunity," whereas article 9(1) speaks of "privileges and immunities" being accorded the "Development Fund for Iraq, its property and assets wherever located and by whomsoever held (including any rights or obligations owed by or to the Development Fund for Iraq). . . ." Proceeds and obligations arising from the sale of oil and gas are not explicitly referenced nor mentioned as such in the context of the article 9(1) grant of privileges and immunities. Given that the DFI is envisioned as the basic mechanism for handling and managing assets derived from the sale of Iraqi oil and gas, in according privileges and immunities to DFI and its property and assets, the language of article 9(1) of British Order 1521 would seem to produce the same effect as that of paragraph 22 of Security Council resolution 1483. The latter grants the proceeds and obligations associated

with the sale of Iraqi-origin oil and gas and the DFI privileges and immunities. The article 9(1) objective of providing privileges and immunities to the DFI and its property and assets seems to be the virtual equivalent. Furthermore, as discussed at length throughout this book, privileges and immunities are completely effective in insulating a recipient from court-requested recognition and enforcement action concerning nonjudicial determinations or decisions rendered by bodies in the same jurisdiction. The only potential question about the language of article 9(1) of British Order 1521 concerns the absence of any reference to "obligations" arising from the sale of oil and gas. However, to the extent that the expression "the Development Fund for Iraq, its *property and assets*" in the article is interpreted as equating "obligations" with "property and assets," then that absence should not prove particularly consequential.

IV. NONJUDICIAL DETERMINATIONS OR DECISIONS PLACED BEFORE COURTS OF A FOREIGN NATION-STATE

The previous section discussed the difficulties that arise from the language of international and national measures that aim to ensure protection of Iraqi oil and gas and proceeds and obligations from their sale from claims by creditors when court recognition and enforcement are sought in the same jurisdiction as that of a body rendering the nonjudicial determination or decision placed before the court. Recalling that discussion, what can be said about the problems associated with recognition and enforcement of such determinations and decisions in a foreign nation-state? In other words, what if the determination or decision of an administrative, arbitral, or other nonjudicial body constituted and operating in France, Germany, or China were placed before a court in the United States, Canada, or Great Britain, and recognition and enforcement were sought because of the presence in those jurisdictions of relevant Iraqi oil and gas assets?

In such a situation, it is clear that observations offered in the preceding sections are important. The the language of paragraph 22 of Security Council resolution 1483 (as amended by resolution 1546 and confirmed and extended by resolution 1905) and the language of various national implementing measures (e.g., U.S. Executive Orders 13303

and 13364, Canadian Iraq Regulations SOR/2004–221, and British Orders 1521 and 1983) do not explicitly speak to the matter of recognition and enforcement. Yet, the want of such clarity aside, terms used in these enactments provide sufficient breadth to reach that matter and preclude recognition and enforcement; and, it would appear to matter little that the particular Iraqi assets targeted were oil and gas or proceeds and obligations from their sale. Also important in connection with requests for foreign-state recognition and enforcement of nonjudicial determinations and decisions are the principles that govern the general field of recognition and enforcement law. It is not within the scope of this study to provide an exhaustive and detailed discussion of recognition and enforcement principles; however, it is useful to set forth the essentials of that law and situate it in the larger context of Security Council resolution 1483 and the relevant U.S., Canadian, and British national implementing measures.

Before addressing these two tasks, however, a few brief comments are in order about the requests for court recognition and enforcement of judicial judgments rendered by the courts of another nation-state. As indicated in the introduction to this chapter, in view of the typical respect accorded by UN member states to the mandates of the Security Council, it is unlikely that many of these situations will surface.[10] Antipathy toward the Council mandates, uncontrollable distaste for the Iraqi regime, or a host of other reasons could precipitate refusal to comply with the directive that member states adopt national implementing measures designed to protect Iraqi oil and gas assets from creditor claims. It is clear that article 25 of the UN Charter directs that UN members adhere to the substantive decisions of the Council, as those decisions create binding international law.[11] Despite this, it is known that pursuant to the domestic constitutional law of certain nations, national law that is later in time and unequivocally diverges from any international law obligation – including one grounded in the UN Charter – leaves that obligation unenforceable in terms of the domestic law of those particular nation-states. From the perspective of insulation of Iraqi oil and gas assets from claims by creditors, that means any such country wishing to scoff at the UN directive established

[10] See text accompanying supra Chpt. 9, Sec. I.
[11] See UN Charter, art. 25.

in Security Council resolution 1483 could adopt legislation that permits judicial action to proceed, with a resulting judgment on such action then being presented to the courts of another nation, and recognition and enforcement requested. Again, these situations are unlikely; however, if they were to occur, the request for foreign-court recognition and enforcement would not only raise many of the general recognition and enforcement law matters described herein, but also would confront the requesting party with certainty that the requested court will respect the measures protecting Iraqi oil and gas assets from effective creditor action.

Now for the essentials regarding the general law of foreign-state recognition and enforcement, and its situation in the larger context of the protection granted Iraqi oil and gas assets by Security Council resolution 1483 and the various national implementing measures. With respect to the bare essentials of recognition and enforcement law, readers familiar with the subject can appreciate that there is a difference in the law between the recognition and enforcement of foreign arbitral determinations or decisions, and the recognition and enforcement of foreign-court judgments. Foreign nonjudicial determinations or decisions – including those resulting from arbitration – comprise the focus herein. However, the essential difference between their recognition and enforcement, and those of foreign-court judgments, is considerable. Concerning foreign-court judgments, few controlling international rules exist; those that do are typically bilateral or regional in character, often relating only to judgments in specific types of disputes. Thus, the matter of recognition and enforcement largely is left to the discretion and standards of each individual nation-state. To illustrate, at one time in the United States, recognition and enforcement of foreign-court judgments hinged on the notions of fair procedures, comity, and reciprocity. In other words, if a foreign-court judgment was settled on in the context of procedural protections deemed fair and impartial, and the courts of the nation issuing the judgment could be shown to recognize and enforce U.S. court judgments placed before them, then a judgment issued by that foreign court would be entitled to recognition and enforcement in the United States.[12] Contemporary U.S. law, however, is based on the idea that recognition and enforcement

[12] See *Hilton v. Guyot*, 159 U.S. 113, 202–203 (1895).

is essentially a matter of the law of each of the individual fifty states in whose courts such might be sought. And, according to commentary accompanying section 481 of the Restatement (Third) of the Foreign Relations Law of the United States, "foreign judgment[s] [are] generally entitled to recognition by courts in the United States to the same extent as a judgment of a court of one State in the courts of another State." This can be undercut if such results in a serious conflict with some interest viewed as essential to the United States. Further, "[a]s in the case of a sister-State judgment, a judgment of a foreign country ordinarily has no greater effect in the United States than in the country where the judgment was rendered. . . ."[13] It is clear that in all cases involving foreign-court judgments against sovereign actors (e.g., the government of Iraq or its instrumentalities), complications that cannot be ignored may exist under provisions of the U.S. Foreign Sovereign Immunities Act (FSIA).[14]

Recognition and enforcement of foreign arbitral awards is quite distinct in being governed by the 1958 New York Convention on the Recognition and Enforcement of Foreign Arbitral Awards (i.e., the New York Convention)[15] and, in cases in which the original dispute involves an investment controversy concerning a nation-state that has been submitted to the World Bank's International Center for the Settlement of Investment Disputes (ICSID), the so-called 1965 Washington Convention on the Settlement of Investment Disputes (i.e., the ICSID Convention).[16] Both international agreements are designed to secure recognition and enforcement of arbitral awards rendered in a jurisdiction, or by an entity, outside or other than that before which recognition and enforcement are requested. Generally, the New York Convention applies to arbitration involving "persons," whether physical or legal, but there are indications that it has also been applied in the context of disputes involving nation-states.[17] Article III of the New York

[13] See Restatement (Third) of the Foreign Relations Law of the United States, sec. 481, comment c.

[14] See 28 U.S.C. Sec. 1330, et seq. See also J. Dellapenna, "Suing Foreign Governments and Their Corporations," 368–395 (1988).

[15] The New York Convention is reprinted at 9 U.S.C.A. Sec. 201.

[16] See ICSID Convention, 575 U.N.T.S. 160 (No. 8359) (1966), 4 *Int'l Legal Mat.* 532 (1965).

[17] See Van den Berg, "Some Recent Problems in the Practice of Enforcement under the New York and ICSID Conventions," 2 *ICSID Rev.* 439 (1987), at Sec. III, "Execution Against a State."

Convention obligates state-parties, under certain exceptions spelled-out in article V, to both recognize and enforce foreign arbitral awards.[18] The ICSID Convention applies very explicitly to situations in which nation-states are involved in an arbitral award. Article 54(1)-(2) of the Convention obligates state-parties to both recognize and enforce ICSID awards.[19] And in deference to measures such as the U.S. FSIA, article 55 provides that "[n]othing in Article 54 shall be construed as derogating from the law in force in any Contracting State relating to immunity of that State or of any foreign State from execution."[20] It bears noting that since 1998 the United States has provided in section 1605(a) of the FSIA for much less immunity for sovereigns than there had been previously.

Regarding the recognition and enforcement law situation in the context of Security Council resolution 1483 (as amended, confirmed, and extended by resolutions 1546, 1859, and 1905) and national measures such as U.S. Executive Orders 13303 and 13364, Canadian Iraq Regulations SOR/2004–221, and British Orders 1521 and 1983, even a foreign arbitral determination or decision otherwise eligible for recognition and enforcement seems to be foreclosed from such. Stated differently, while the terms of the New York and the ICSID Conventions speak of an obligation of courts of signatory countries to recognize and enforce foreign arbitral awards, the purpose of paragraph 22 of resolution 1483 is to insulate Iraqi oil and gas assets from the kind of jeopardy to which they may be subjected by deserving creditors. Even in nations where the resolution's national implementing measures are written so that nonjudicial actions may proceed when original judicial actions are prohibited, subsequent judicial action to convert determinations or decisions resulting from the permitted nonjudicial action into meaningful and strong remedies is precluded. Original as well as subsequent court actions to secure recognition and enforcement of original nonjudicial action are prohibited. The very same language of the relevant measures serving to work that result also preclude court recognition and enforcement of foreign nonjudicial arbitral awards. Attention should be called to the fact that the recognition and enforcement obligations of the New York and the ICSID Conventions face

[18] See New York Convention, supra note 15 at arts. III and V.

[19] See ICSID Convention, supra note 16 at art. 54(1)-(2).

[20] See id. at art. 55.

complications when an administrative or another form of nonarbitral, nonjudicial determination or decision is involved. Nevertheless, were those to be seen as somehow eligible for recognition and enforcement, the terms of the controlling law protecting Iraqi oil and gas assets also would foreclose court recognition and enforcement of the foreign awards.

V. CONCLUSION

Consonant with the preceding section's outline of the general principles regarding recognition and enforcement of foreign-state nonjudicial (and judicial) determinations or decisions, it must be emphasized that the focus not only of this chapter, but also this entire study is on particular Security Council resolutions, Iraqi law, and national implementing measures of other nations and what is said about protecting Iraq's oil and gas assets from claims by creditors. There is no question that if a foreign-state nonjudicial (or judicial) determination or decision were presented in a court in the United States, Canada, Great Britain, or elsewhere and recognition and enforcement were requested, the recognition and enforcement principles and numerous related complexities would require detailed and in-depth consideration and exploration. The intricacies of those principles and their affiliated legal rules are deliberately avoided in this study in order to not lose concentration on the important subjects examined herein. Much has been written by skilled and expert commentators about the law of recognition and enforcement. By conscious choice, the narrow rules specifically addressing protection of Iraqi oil and gas assets comprise the touchstone throughout the discussion that proceeded. This focus is maintained not only to provide sedulous review of a previously unanalyzed and neglected body of law, but also because the nuances and angles relative to the general law of recognition and enforcement would considerably expand this study and result in an examination of a subject more appropriately left to the savants of that field of law.

10 LESSONS LEARNED FROM THE IRAQI EXPERIENCE TRANSFERABLE TO OTHER INSTANCES OF DEBT OWED BY RESOURCE-RICH NATIONS

I. INTRODUCTION

When determining the lessons learned from the experience of insulating Iraqi oil and gas assets from claims by creditors – and the extent to which those lessons are transferable to other instances of debt owed by resource-rich nations – it is imperative to be aware of the enormous diversity of situations and circumstances that may exist. There are basic lessons from the Iraqi case that are generally applicable to a host of other comparable cases. Comment will be made on many of these in the pages that follow in this chapter. At the same time, however, recall that the Iraqi situation is somewhat unique.

For starters, the resource with which Iraq is rich is essential to the functioning of the fundamental elements of modern society. At some point in the near future, there is likely to be a shift toward greater dependence on renewable sources of energy to power the wheels of commerce and transportation. At present, the world remains largely dependent on fossil fuels and, when it comes to assuring rapid personal mobility, the dependence on such is total. Additionally, Iraq is situated in one of the world's principal international political hotbeds. The Arab–Israeli crisis has confounded and bedeviled the world community for the past half-century. There is little reason to believe that the inveterate enmities precipitated by the dispossession of many Palestinians coincident with the founding of the State of Israel, the sheer antipathy

engendered by the several Middle East wars and Israeli expansions, and the well-deserved suspicion spawned by both the numerous attacks on innocent Israeli citizens as well as the vitriolic diatribe spewed constantly by radical elements ill-advised on "how to make friends and influence people" are likely to suggest any immediate resolution to the problem. Closely related to Iraq's physical location in the Middle East is its temporal connection to the al-Qaida attacks on the United States on September 11, 2001, and the so-called war on terrorism. This is not to suggest anything more than that the problem of Saddam Hussein and the difficulties associated with allied involvement in Iraq after his removal existed in a milieu of reordering relations between the West and the Islamic world. Against this backdrop, the centrality of Iraq remains for many an article of faith to this very day.

It is obvious that in other situations, the resource with which a particular debtor nation is rich may be substantially different. The resource may be one that is important, but not nearly as essential as oil and gas (e.g., chromium or molybdenum for hardening steel). Or it may be a resource that is highly desired, but largely dispensable for modern life (e.g., gold or diamonds). Moreover, the resource may be one to which the creditor nations themselves have ample access by virtue of deposits within their own territory, or in the territory of others with whom they have excellent relations. Likewise, the survival of the government (or ruling political philosophy) of the debtor nation may be perceived as not nearly as important as that of the democratically elected government of Iraq. This could be a function of where the concerned debtor nation is geographically located, the gravamen of its persona in the world community, and the degree to which the debtor has a role to play in important international political issues. Furthermore, the importance to the debtor nation of the particular resource in which it is rich has its own significance. That is to say, in a country like Iraq where 95 percent of foreign revenue earnings are produced by export sales of oil, the situation is entirely different from the case of another country that produces oil for the world community, but substantially supplements foreign-export earnings through shipments of many other items. Finally, the financial position of the creditor nations, or those to whom debt is owed, may be quite distinct from what exists relative to Iraq. Typically, it is large capital-exporting nations (or their nationals) that are owed debt. However, this would seem to cut both

ways: such nations (or nationals) are more capable of absorbing losses or delays associated with debtor insulation; they also are more capable of leveraging their capital-exporting power into pressure to insist on debt payment.

The interactions and relationships among these and other factors can make for a complex and interesting mix in the context of addressing any particular international-debt situation. Iraq may submit to a specific type of analysis under these factors, whereas another nation may find itself in an entirely distinct position. Readers should be aware of this as the subsequent sections attempt to divine the important lessons learned from the various measures – international and domestic – that establish the debt-relief regime for Iraq's financial obligations, and offer protection to Iraqi oil and gas assets. Other than concerns with the legal regime protecting Iraqi oil and gas assets from claims by creditors, what may make sense in the context of Iraq, may not make any sense at all in the context of another country.

II. CLAIMS-RESOLUTION MECHANISMS AND DEBTOR-NATION LAWS PERMITTING CONTRACT ALTERATION AND ESCAPE

As indicated in Chapter 1, various mechanisms have been used in the Iraqi situation to address that country's debt load, including the UNCC, the IDRO, the Paris Club, and others.[1] It seems clear that some of the mechanisms were deliberately designed to deal with claims that arose from specific situations (e.g., the First Gulf War). Others handled claims held by particular nation-states, typically those with membership in a certain formal or informal organization or group. Still others dealt with claims from entities of a specific type, such as banking or financial institutions. To the extent that there has been use of mechanisms aimed at providing claim resolution involving smaller, private entities, participation has been largely voluntary.

At the same time, Iraqi governing institutions – principally because of dissatisfaction with some oil- and gas-development agreements concluded between Iraqi authorities and foreign entities subsequent to the First Gulf War – have seen fit to legally permit the unwinding of those

[1] See supra Chpt. 1.

agreements. As discussed in Chapter 4, both article 141 of the 2005 Iraqi Constitution[2] and article 54 of the 2007 KRG oil and gas law (No. 22)[3] speak to this matter. There may be other constitutional provisions that can be raised as limitations on these permissions; nevertheless, any authority that empowers governing entities to nullify and void or refuse to honor, or imposes amendments on preexisting contractual commitments, simply compounds the potential debt load borne by the acting nation.

With respect to claim-resolution mechanisms, it would seem that the more comprehensive the mechanism is in reaching the most encompassing range of claimants, the better. After all, does it make sense to link a series of debt-reconciliation mechanisms with a temporary prohibition on legal action regarding such debt, and then structure the reconciliation mechanisms so as not to accommodate the full panoply of potential creditors? By conjoining debt reconciliation with a prohibition on legal action, the objective apparently is to drive creditors away from standard legal measures aimed at relief and toward reconciliation through the prescribed mechanisms. However, if the mechanisms for debt reconciliation fail to bring in all potential creditors, those remaining may be precluded from pursuing relief through standard legal action as long as the prohibition is in effect. Once the prohibition is lifted, their unrequited claims will generate all the protracted and expensive proceedings that structured reconciliation was designed to avoid. Therefore, it is imperative that any effort to structure debt-reconciliation mechanisms reflects on the entire spectrum of potential claimants, which runs the gamut from foreign nations and their instrumentalities and agencies; regional and international governmental development and financial entities (e.g., regional development banks, the IMF, and the World Bank); major private commercial lenders, product sellers, and service providers; intermediate and smaller lenders, product sellers, and service providers; and individual operators and entrepreneurs that have contractual arrangements with the debtor nation. Additionally, the effort to establish debt-reconciliation mechanisms should consider making the use of them the sole method for creditors to gain satisfaction on obligations owed.

[2] For art. 141 of the Iraqi Constitution, see text accompanying supra Chpt. 4, note 3.
[3] For art. 54 of the KRG law, see text accompanying supra Chpt. 4, notes 4 and 5.

In some situations, there may be reluctance on the part of relevant governments to agree to participate caused by a philosophical aversion to anything less than the use of standard legal mechanisms. Also, reluctance may similarly derive from pressure of affected claimants dissatisfied with the possibility that claims resolved through the debt-reconciliation mechanism may have to be settled for a small percentage of what actually is owed. Yet, any situation in which some opt out of participating could result in those not in the system ultimately receiving a larger percentage of the original claimed amount. This undermines the interest of claimants in any future cases being inclined to support the establishment of a comprehensive debt-reconciliation mechanism.

The Iraqi situation involved various efforts at debt reconciliation, mounted in many cases by distinct groupings of entities with common interests. The Paris Club states aimed to assist the fledgling democracy in Iraq by forgiving much of the debt owed to those states.[4] This was perceived as freeing up future oil and gas export-sales revenues for the reconstruction and development of Iraq. The UNCC concentrated on First Gulf War claims[5] – which was, in part, likely retributive for Saddam Hussein's precipitation of that war. The efforts by the IDRO endeavored to entice many of the claimants with intermediate and smaller size monetary demands, but the voluntary nature of participation has not been able to exhaust all remaining claimants.[6] To a certain extent, benefits of a unitary claim-reconciliation structure are lost whenever satisfaction of debts is addressed through multiple structures. Multiple reconciliation structures can result in disparate treatment of creditors who have claims with striking similarities. Rationales and principles relied on by reconciliation bodies may not appear to be shared and uniform. In short, seemingly inexplicable divergences can surface whenever multiple debt-reconciliation structures are utilized. Divergences always raise suspicion of partiality or bias – whether or not warranted – and are detrimental to public confidence in the neutrality and essential fairness of the involved decision-making entities.

With respect to debtor-nation legal efforts to permit the unwinding of certain types of preexisting contractual commitments, the transitory

[4] See text accompanying supra Chpt. 1, Sec. I.
[5] See text accompanying supra Chpt. 1, Sec. II.
[6] See text accompanying supra Chpt. 1, Sec. V, especially notes 59–75.

nature of governing regimes in many countries – and, unfortunately, this is often the case in many that are resource-rich – suggests an ever-present inclination to take that approach. New regimes replace old regimes because of strongly held differences in perspective about how certain matters should be handled. Operating in that context, it is quite possible that contractual arrangements of predecessor regimes – although negotiated and entered into under the most solemn of circumstances – can be deemed unworthy of adherence, with legislative or regulatory action taken to permit those arrangements to be escaped. Article 141 of the 2005 Iraqi Constitution, and article 54 of the 2007 KRG oil and gas law (No. 22), both seem to fall into that category. Distinct from strategies that might be used to minimize the chances of a nation disregarding preexisting commitments, any legislative or regulatory action taken to nullify or void, or to ignore and rewrite the terms of earlier contractual arrangements should be absolutely clear. Recall from the discussion in Chapters 4, 5, and 6, in particular, that untold numbers of ambiguities are associated with the language of those and other relevant Iraqi provisions. In any event, however, if the legislative or regulatory architects of a nation's new regime desire to permit departure from past contractual commitments – whether involving oil and gas, natural resources, or sundry other matters – they should take great care to be as specific and detailed as possible. Unless unclear in their own approach as to what such legislative or regulatory permission should extend, the measures implemented to effect that desire should precisely and exhaustively define (1) the exact nature of each and every commitment that can be voided or rewritten (e.g., particular forms of contracts recognized in certain industries – e.g., PSCs, TSAs, or TSCs – and other more general types of agreement); (2) the parties who might hold such commitments (e.g., those who contracted with the debtor nation now interested in altering its commitment, or third parties that contracted with a principal contractor); (3) the subject matter that the commitments involve (e.g., exploration for and production of oil and gas, the performance of collateral activities associated with oil and gas production); and (4) the specific time during which a potentially affected commitment was to have been negotiated. Without such precision, those involved – from either the standpoint of being a contracting party or asked to rule on the applicability of legislation or regulations aimed at undoing a contractual

commitment – are asked to enter the netherworld of interpretive speculation.

It should also be noted that both article 54 of the KRG oil and gas law (No. 22), and article 141 of the Iraqi Constitution pose an interesting question when juxtaposed with paragraph 22 of Security Council resolution 1483,[7] and the various national implementing measures designed to protect Iraqi oil and gas assets from recovery actions by creditors.[8] Specifically, is it possible to view articles 54 and 141 as somehow indicating a permissive attitude toward claim actions by creditors, and thereby understand them as informing the meaning of the UN and national measures that seem to preclude creditor recovery efforts? It is clear that paragraph 22 of resolution 1483 and the various national implementing measures endeavor to insulate Iraqi oil and gas assets from legal actions by creditors. Nonetheless, the language of the pronouncements is subject to interpretation on questions such as whether they apply to only court proceedings, or also to administrative, arbitral, and other legal actions. Given that article 141 of the Iraqi Constitution permits the KRG to exercise the authority to undo some oil and gas development agreements, and article 54 of the KRG oil and gas law authorizes administrative determinations regarding whether some of the agreements are to be revised and reworked, is it at all possible to conceptualize the authorization and KRG effort at revision and reworking as informing the scope of the UN and national measures to protect Iraqi oil and gas assets? The simple answer is a resounding "no." It would be difficult to place much credence in the ability of national Iraqi measures on the undoing of oil and gas contracts to cast light on the meaning of international and foreign measures to protect Iraqi oil and gas assets from claims by creditors. The major effect of articles 54 and 141 is to potentially increase Iraq's debt load by facilitating contract departures that are sure to generate claim actions. Beyond the inability of such Iraqi measures to inform the meaning of relevant international and foreign standards, however, articles 54 and 141 stand in stark contradistinction to the UN and foreign standards that protect Iraqi oil and gas assets. The former speak to the permissible "undoing" of contracts, and the latter to

[7] For the content of para. 22, see supra Chpt. 3, note 21.
[8] See, e.g., supra Chpt. 8.

the protection of Iraqi oil and gas assets from claims by creditors. Clearly, it may prove useful in future situations where interpretive guidance is sought from the interface between legal measures bearing some relation to one another to provide in explicit terms the precise extent of that relationship. In the case at hand, this reemphasizes the value of international and national implementing standards that spell out the exact nature of the proceedings and processes foreclosed to creditors.

III. USE OF INTERNATIONAL INSTRUMENTS TO ARTICULATE STANDARDS

Both Security Council resolutions and international status of forces agreements have addressed treatment of the debt load owed by the Iraqi government to foreign creditors. Of central importance is the language of paragraph 22 in resolution 1483 (as supplemented by resolution 1546[9] and extended by resolutions 1859 and 1905).[10] Working in conjunction with these resolutions, and serving as part of the foundation for resolution 1859, is the U.S.–Iraq SOFA of December 2008,[11] and in particular its paragraph 26. The development of the specific terms of paragraph 22 of resolution 1483 is examined in Chapter 3, and the details associated with carefully parsing that language are discussed in Chapters 7 and 8. Several lessons can be learned from the efforts of paragraph 22 to protect Iraqi oil and gas assets from claims by creditors.

The first lesson concerns the nature of the proceedings from which protection has been accorded. As seen, the language of paragraph 22 references "legal proceedings," but it is understandable that the reach of that reference may be confusing. It is preferable to lean toward greater specificity in identifying the nature or character of the proceedings or processes from which a resource or resource-related asset held by a debtor nation is to be immune. The follow-on references to "attachment, garnishment, or execution" are helpful; yet, the language of the May 9 and May 15 drafts of what became paragraph 22 of

9 See supra Chpt. 3, Sec. IV.

10 See id., Sec. V.

11 See supra Chpt. 7, Sec. VII, especially notes 43–46.

resolution 1483 are even more helpful.[12] The point is that if the intent is to insulate particular assets (e.g., Iraqi oil and gas and proceeds and obligations from their sale) from every form of court, administrative, arbitral, or other action to determine legal rights associated with debt owed to creditors – indeed, if the intent is to extend such protection to include legislative or executive action or measures as well – then that intent should be very carefully and completely spelled out. In the absence of doing so, there always will be questions about whether such was intended. True, it may be that, under the most cribbed interpretation of the words "legal proceedings," only court action is foreclosed; nevertheless, noncourt action that permissibly goes forward can eventuate in nothing since subsequent court action to put "teeth" into an arbitral award, for instance, is precluded as well. As previously noted, costs and inconveniences are associated with defending noncourt proceedings. If an international standard seeks to protect a debtor nation's assets from having to absorb those costs, it is preferable to employ a considerably more encompassing formulation than "legal proceedings."

The second lesson arising from paragraph 22 of Security Council resolution 1483 concerns the assets to be protected. Paragraph 22 speaks in terms of "petroleum, petroleum products, and natural gas," and "proceeds and obligations" arising from sales thereof (and the DFI). Formulations of this type can raise questions about whether only lifted, or even inground oil and gas, or mere fractional-share interests in either, are included within the protection intended. Intense consideration must be given in advance to the precise nature or character of the assets protected; general references to broad classes or categories of assets can prove problematic and must be avoided. Specificity and exactitude are the watchwords. It makes little sense to offer language formulations that reasonably can be construed as including only a certain asset if a more inclusive and wide-ranging group of assets were intended. If petroleum and natural gas are to be immune from creditor action – whether the petroleum and natural gas is *in situ* or in a production-field holding-tank, a pipeline, a storage facility, or a shipping or transport vessel – it should be clearly stated. If interests in the entirety of certain oil or gas reservoirs or what has been produced from them, as well as interests in fractional shares of such, are to be

[12] For the language of those drafts, see supra Chpt. 3, Sec. III, especially notes 31–34.

immune, that should be clearly stated. Indeed, if any asset's importance to a debtor nation is thought to be of sufficient magnitude to warrant consideration for protection, then it should be set forth with painstaking clarity.

Paragraph 22 of Security Council resolution 1483 basically is cast in language that focuses on the Iraqi assets that are insulated from prosecution. With the exception of the paragraph's phraseology dealing with the lifting of immunity once title has passed to an initial purchaser, the thrust essentially indicates that it is Iraqi oil and gas assets that are protected from creditor action. This leads to a third lesson: Other than an initial purchaser, if it does or does not matter who holds or has a right recognized by law in a debtor nation's protected assets, then this needs to be explicitly indicated. It is reasonable to expect that a debtor nation is protected from action that targets an insulated asset. To the extent that it is thought vital to protect third parties who may hold or share interests in protected assets with the debtor nation, it should be provided for in unequivocal language. A complication with the approach of paragraph 22 is that it leaves open whether Iraqi-origin oil and gas assets in the hands of third parties that have not acted as a classic purchaser should remain within the ambit of the resolution's basic protection. Barring an actual transfer of legal title through a purchase, if it is considered essential to protect certain debtor-nation assets from creditor action regardless of who holds an interest therein, then the language of the relevant protective measure should make that intent manifest.

A fourth lesson derived from that same paragraph has to do with the protection of "proceeds and obligations" arising from the sale of Iraqi oil and gas, and the affiliated matter of protecting fractional-share interests in proceeds and obligations, or oil and gas proper. What is the exact objective in protecting "obligations" arising from the sale of oil and gas? Is it to protect every imaginable right recognized by law and generated as a consequence of the sale of petroleum, petroleum products, or natural gas? Or is it an attempt to provide protection from creditor action only with respect to the legal responsibilities or duties created by a contract to sell petroleum, petroleum products, or natural gas? It will be recalled from Chapter 8 that U.S. Executive Order 13303 speaks of protecting proceeds, obligations, "or any financial instruments of any nature whatsoever" that arise from the sale of

Iraqi oil or gas.[13] Is this an indication that the scope of the language in paragraph 22 of Security Council resolution 1483, in protecting oil- and gas-sale "obligations" from creditor action, aims at protecting far more than the mere responsibilities and duties created by a contract to sell them? Again, there is no substitute for absolute clarity. The meaning of the reference in paragraph 22 to "proceeds" may be somewhat understandable; it would certainly help relative to "obligations" if more attention had been given to specifying exactly what was meant. Interpretive arguments can be advanced suggesting the breadth of that term. However, if the drafters had aspired to suggest that all manner of rights recognized by law and arising from the sale of oil and gas were to be eligible for protection from legal proceedings, then it would have been significantly better to sedulously detail that intent with additional illustrative language and all-inclusive phrases. Had their intent been to the contrary – that is, to confine what was protected to the rights emerging from oil and gas sale contracts – then that also should have been spelled out in some detail. It is always unsatisfactory to rely on terms that are fraught with ambiguity. Specificity and detail are the best prescription for legal drafters. These two virtues serve to preempt potential disputes and minimize the chances of any controversy ever maturing.

IV. TIGHTENING THE STANDARDS OF NATIONAL IMPLEMENTING MEASURES

As just alluded to, Executive Order 13303 (as elaborated by Executive Order 13364[14] and extended by presidential Notice of May 20, 2009[15]) can be contrasted with the language of paragraph 22 of resolution 1483 in various respects. It is also one of the many national implementing measures adopted to execute the protection objectives of that paragraph. The national implementing measures also reveal problems that comprise the basis for lessons applicable to future situations involving resource-rich debtor nations that may need the revenues to ensure their rehabilitation, economic growth, and development.

[13] See text accompanying supra Chpt. 8, Sec. II.

[14] See id.

[15] See supra Chpt. 8, Sec. I, especially notes 2–4.

Reflecting on the implementing national measures of the countries examined in Chapter 8, at least five firm lessons emerge. In general, efforts to convert these lessons into application focus on tightening the standards provided in the national implementing measures. The first two lessons concern over-reaching aspects of the measures, and under-reaching. An example of over-reaching is the language of section 1 of Executive Order 13303, which addresses immunity from judicial process for not only Iraqi-origin oil and gas, and the proceeds and obligations arising from their sale, but also for proceeds and obligations "related to the sale or marketing" of oil and gas. The language of paragraph 22 of resolution 1483, by comparison, references protection only for oil and gas, and the proceeds and obligations from their sale (and the DFI). Thus, the inclusion in section 1 of Executive Order 13303 of protection for both proceeds and obligations arising from the sale of oil and gas, and the proceeds and obligations "related to the sale or marketing" of such, precludes creditor action against Iraqi proceeds and obligations generated by either "marketing" (as distinct from sales) activity, or by activities merely "related to" sales (or marketing), rather than integrally involving actual sales. In other words, the way the language of section 1 is written, the simple fact of marketing Iraqi-origin oil and gas may produce proceeds and obligations beneficial to Iraq. It does not matter that the marketing activity eventuates in the sale of such, since the language of the Executive Order would entitle what has been produced by the "marketing" itself to be immune from judicial process. Likewise with respect to proceeds and obligations stemming not from integral components of oil and gas sales (or marketing), but rather from matters characterized as no more than "related to" oil and gas sales (or marketing). For example, revenues paid directly or indirectly to Iraq by a purchaser of Iraqi oil and gas, or the obligations emerging from the contract involving such, are clearly within the notion of proceeds and obligations "arising from" the sale. The notion of proceeds and obligations "related to" such a sale, however, appears far more encompassing and capable of placing under the protective insulation of the Executive Order considerably more than what was contemplated by the foundational Security Council resolution. It may be that individual nation-states are entitled to decide precisely what they desire to designate as eligible for protection from creditor action within their legal system. Clearly, however, if the desire reaches beyond

what was designated by any foundational UN resolutions to be protected, it is clear that for that nation-state to avoid claims of acting *ultra vires*, it would be compelled to be able to point to legal authority other than the relevant resolution adopted by the Security Council.

The opposite phenomenon of under-reaching – that is to say, providing protection against creditor claims for fewer assets of a particular debtor nation than contemplated in a relevant international directive – is illustrated in the reference in section 1 of Executive Order 13303 that Iraqi oil and gas assets are immune from "judicial process." As suggested in Chapter 8, this differs from the language of Security Council resolution 1483, which speaks of "legal proceedings," thereby raising the question of interpretive distinctions. Admittedly, the idea of "judicial process" is subject to various constructions; it is not beyond the realm of reasonableness to view that concept as confined to process before a judicial-branch organ or instrumentality (i.e., a court). To the extent that the nature of the immunity provided under section 1 of the Executive Order is so limited, it appears that the section fails to extend to all legal proceedings as mandated by the controlling resolution. As previously discussed, protection of Iraqi oil and gas assets from legal proceedings can prevent their subjection to any proceeding designed to determine or settle legal rights and obligations in respect to them. Precluding only judicial process (i.e., court process) therefore fails to fully effectuate the relevant Security Council obligation as set forth in paragraph 22 of resolution 1483. Such under-reaching can be problematic to the extent that it results in different countries going their own way; it also can stir controversy within the UN itself. To avoid this, when the Security Council issues directives, the language of the relevant resolutions should be absolutely clear and leave no "wiggle room" for member states. Directives should be explicit, and they should provide that member states have no latitude to stray or vary from the terms provided. In the context of potential creditor claims against resource-rich debtor nations, this may have the effect of freeing those nations from otherwise inconvenient processes they might have to endure.

In addition to the lessons concerning over-reaching and under-reaching, a third lesson is associated with how both Executive Orders 13303 and 13364 provide rather lengthy detail regarding the types of judicial process that are prohibited, and the extent of the protection

provided to the CBI. With respect to prohibited judicial processes, recall that paragraph 22 of Security Council resolution 1483 refers to attachment, garnishment, or execution. The chapeau of section 1 of Executive Order 13303, on the other hand, refers to not only those three, but also "judgment[s], decree[s], [and] lien[s]" as well. This emphasizes the idea that any judicial process – not only attachments, garnishments, and executions – is precluded against Iraqi oil and gas assets. As observed earlier, the idea that any listing is just illustrative is further supplemented by the fact that the listing in section 1 is followed by the statement that what is prohibited with respect to Iraq is any "other" judicial processes besides those listed. Whether the concern is with the language of Security Council resolutions, or implementing national measures, the more detailed the better. That same kind of detail also exists with respect to the language added to the U.S. approach by Executive Order 13364's insulation of the CBI. As pointed out in Chapter 8, one accomplishment of Executive Order 13364 was the dividing of section 1 of Executive Order 13303 into subsections (a) and (b): the former having part (i) concerning the DFI; part (ii) concerning Iraqi petroleum, petroleum products, natural gas, and proceeds and obligations; and part (iii) concerning the CBI. In regard to the new section 1(a)(iii), the CBI and its assets are provided protection. With attractive specificity, it speaks of CBI "accounts, assets, investment, or any other property of any kind, owned by, belonging to, or held by the Central Bank of Iraq, or held, maintained, or otherwise controlled by any financial institution of any kind in the name of, on behalf of, or otherwise by the Central Bank of Iraq." In looking for a paradigm of detail and completeness, we would do well to consult the language added by Executive Order 13364.

A fourth lesson from the Iraqi oil and gas asset-protection effort concerns the confusion regarding legal standards that is injected by national implementing measures that depart from what the relevant international rules provide – in this case, those established by resolution 1483. Clear differences exist between a grant of privileges and immunities, and a grant of immunity alone. Paragraph 22 of resolution 1483 declares that Iraqi oil and gas have immunity from legal proceedings, and the proceeds and obligations associated with their sale (and the DFI) have privileges and immunities equivalent to the UN. By way of contrast, recall that British Order 1521 provides in article 9 for

privileges and immunities for the DFI, and in article 10 for straight immunity for Iraqi-origin petroleum, petroleum products, and natural gas.[16] The British Order makes no clear reference to what type of protection can be expected by proceeds and obligations arising from the sale of oil and gas. Given that the DFI collects and manages oil and gas revenues, arguments may be advanced that such come within the ambit of the DFI's protection; however, the absence of any explicit reference to proceeds and obligations in the context of British Order 1521's mention of privileges and immunities raises questions about their protection. Obviously, when it comes to the need for national implementation of international standards, there is always potential for varying approaches, thus presenting the prospect of every variance being deemed a true departure. Although it is highly unlikely that the language of British Order 1521 is intended to signify this type of objective, there is no doubt that questions regarding intentional departure are raised. Confusion of this nature is not especially helpful; to avoid it, therefore, if national implementing measures seek only to render faithful effectuation of international rules, they should assiduously refrain from linguistic formulations that confuse their intention. In the event the desire is to provide something different from what the international rules provide, language that avoids confusion on that point should be employed. Likewise, if the desire is to replicate what the international rules provide, then language that makes that clear should not be pushed aside. This practice avoids all confusion, makes life far easier for those required to work with the standards, and puts the decision-maker in a position of greater confidence when confronted with disputes regarding compliance.

The final lesson associated with tightening the standards in national implementing measures has to do with the fact that national measures may sometimes make the error of equating certain matters intended by the foundational international rules to be entirely distinct. A prime illustration of this can be found in Canada's Iraq Regulations SOR/2004–221 indicating that "no legal proceeding . . . shall lie" against Iraqi-origin oil and gas, or the proceeds and obligations arising from their sale. The idea is to place Iraqi petroleum, petroleum products, and natural gas, as well as the proceeds and obligations associated with their

[16] See supra Chpt. 8, Sec. VI, especially notes 29–31.

sale, under the exact same type of protection.[17] But as is plain from the language of paragraph 22 of Security Council resolution 1483, proceeds and obligations are under a different and more protective degree of insulation than are simple Iraqi oil and gas. The Canadian Iraq Regulations, however, must view privileges and immunities as tantamount to the immunity to which Iraqi oil and gas are entitled; otherwise, it is difficult to explain why the regulations treat oil and gas, and the proceeds and obligations, exactly the same way. As discussed herein, language used in national implementing measures must avoid confusion about whether the intent is to depart from what is called for in foundational international rules. Similarly, language in national implementing measures that equates matters that the foundational international rules view as entirely distinct also must be avoided. There can be no controversy that a distinction exists between immunity for oil and gas proper, and the privileges and immunities for proceeds and obligations derived from their sale. It is a mistake for national implementing measures to contain language that does not reflect this distinction. National implementing measures should not equate matters that deserve unique and different treatment.

V. CONCLUSION

Rather than summarize the observations recounted in this chapter, three additional comments are offered here as possible lessons from the protection of Iraqi oil and gas assets through international and national implementing measures. The comments concern broader lessons that emerge from the attempt to insulate Iraq from claims by creditors.

One of these broader lessons concerns the duration of protection from claims by creditors. The way the situation has been handled with respect to Iraq has resulted in the duration being matched to the preferences prevailing in the Security Council on that matter at any given time. In recent years, the Security Council has voted annually to reestablish protection for an additional consecutive twelve months, whereas current Security Council resolution 1905 provides for expiration of the protection at the end of 2010. Given that Iraq is still a

[17] On the Canadian Iraq Regulation, see id., especially notes 32–34.

long way from being stable and flush with cash, it is reasonable to expect future extensions of the oil and gas asset protection. From the perspective of those owed money by Baghdad, this situation is not entirely satisfactory. Clearly, they would like to know that, at some point in time, they can press their claims before an appropriate legal forum. As it stands now, however, no one can precisely say when that might be – perhaps in a year or two or, based on the deterioration in Iraq's security situation or the price of oil on the international market, several more years in the future. An unfortunate aspect of this situation is that without a fixed expiry of protection from creditor claims, incentive for Iraqis to do what is necessary to arrive at an accommodation between contending ethnic and sectarian factions is dampened. In other words, why work toward resolution of the many civil problems facing the nation, if an unhealthy situation militates in favor of extending protection of oil and gas assets from legal action? Indeed, although some may disagree, it seems a case can be made for the proposition that the entire array of policies associated with Iraq functions as a disincentive for that nation to move toward true and lasting self-sufficiency.

Another related lesson concerns the possibility of an international effort to move in the direction of somehow structuring an international legal regime that includes the basic features of the theory of odious debt (see Chapter 2). Quite possibly, the Iraqi situation may be seen by those hesitant to subject resource-rich but economically disadvantaged or politically dysfunctional and fiscally unsound nations to the claims of creditors from foreign countries as a watershed event in the movement toward an international legal regime of wide-ranging debt forgiveness. Admittedly, the theory of odious debt is limited; nevertheless, it aims to exculpate debtor nations from certain types of claims by creditors. Likewise, the debt-relief efforts undertaken by the Paris Club, the IDRO, and others also seek debt minimization or elimination. Clearly, a modicum of sentiment exists for insulating nations situated in certain positions from preexisting debt when a significant change in such nations' governing philosophy occurs. Whether the critical mass of altered sentiment exists to move the international community in the direction of a legal paradigm of debt forgiveness is another question. In the event that such can be found, however, perhaps the Iraqi situation can be used to catalyze that sentiment into a formal process that results

in a rule of international law officially protecting such debtor nations from claims by all or certain creditors. In addition to putting specific potential creditors on notice of the fact that international law may not recognize debt they could be willing to extend to certain nation-states, it would have the effect of eliminating the need to address the question of insulation from debt on a case-by-case basis. There are considerable difficulties associated with formulating a specific international legal rule on the matter. Consequently, given the complicated and controversial matters that daily confront the community of nations, it is not certain that there is a willingness to do more than address debtor-nation protection from creditor claims on an ad hoc, as-needed, and individual basis.

The Iraqi debt situation also prompts comment on political-risk insurance – that is, insurance obtained by creditors of Iraq at the time of or immediately after contracting, and designed to insure against any losses subsequently incurred as a result of Iraqi government nonperformance. Where used, it must be acknowledged that such political-risk insurance merely shifts the burden of making claims against Iraq from the principal creditor to the insurance issuer that is subrogated to the creditor's claim. In many instances, however, such insurance may not be obtained, thus leaving the creditor who contracted with Iraq in the position of having to use whatever measures are available to obtain satisfaction for the obligation or debt it holds. From the perspective of those who contracted with Iraq and now are in the unenviable position of enduring Iraqi nonperformance, political-risk insurance is distinctively attractive. Depending on the specific terms of the insurance contract, the beneficiary can rest assured that it will receive remuneration of the type expected when it originally contracted with Iraq, either because of an Iraqi decision to comply with its contractual obligations, or because Iraqi nonperformance will result in payment according to the terms of the insurance contract itself. The Iraqi-debt situation, which resulted from the failure to fulfill contractual commitments, suggests that in the future those who loan money to, contract to perform services for, or supply goods designed to meet the needs of sovereign nation-states or their agencies or instrumentalities would be well advised to consider the risks of nonperformance associated with such contracts, and to weigh the need for a political-risk insurance policy that can compensate for default on or repudiation of

such contracts. Again, it is clear that with respect to the matter of creditor claims against debtor nations like Iraq, the effect of such insurance is plainly to shift the burden from the principal who contracts with the debtor nation. Nonetheless, from the standpoint of the principal, the insurance has the beneficial effect of allowing the principal to receive compensation expected as a result of the loan or the service or goods contract. The fact that international standards precluding action for claims by creditors are foreclosed, or the fact that national implementing measures designed to produce the same result are adopted, do nothing to obstruct the receipt of expected payments. To the extent that political-risk insurance was not used as often as it should have been in the Iraqi situation, its attractiveness in comparable or analogous situations in the future seems a palpable lesson.

A final comment is in order with respect to the entire matter of debt relief or forgiveness. That comment has to do with the fact it may just be possible little true sentiment exists in the community of nations for such relief or forgiveness. Indeed, although a minority of nations may be vociferous in voicing the sentiment, the vast majority may be disposed to prevent the development and adoption of any general international legal principle aimed at universality of debt relief or forgiveness for newly-formed governing structures that displaced earlier regimes that had incurred debt for offensive or suspect purposes. If this more accurately reflects the mood of the international community, then it is obvious that the Iraqi situation teaches the need for great and enduring circumspection when it comes to contracting with other nations or their agencies or instrumentalities. It may seem that the contract entered into supplies one with all the necessary requisites to permit essential action to receive satisfactory compensation in the event of breach, repudiation, default, or nonperformance. However, through measures adopted by the international community and implemented nationally, such action may be foreclosed. Inherent in this is the lesson to be extremely circumspect and cautious about contracting with governmental regimes that display certain attributes, regardless of the fact that they exercise their authority over nations that are resource-rich. Although there may be an expectation of either contractual performance or necessary action to secure compensation in the event of nonperformance, the fiscal and economic needs of successor regimes could lead the international community or individual states to bar access to

standard procedures available to creditors seeking claim satisfaction. In choosing a course that seems unsympathetic to deserving creditors, the objective may be to drive home the point that certain regimes should not be dealt with – and, in the event that they are, successor regimes have a prior and superior claim against that nation's resources and assets.

EPILOGUE

The estimated $50 billion to $70 billion in debt currently owed by Iraq to foreign creditors, as reduced by the late 2009 Chinese deal of debt for oil, and other possible reductions not widely reported in the news media, eventually may be repaid, negotiated downward, entirely forgiven, or remain the subject of controversy and, ultimately, various legal actions. The objective of this study was to survey the background regarding the nature of that debt, the contentions that it is legally escapable under the theory of odious debt, and the provisions of Iraqi law – adopted by both the federal government and the KRG – that could precipitate additional claims as a consequence of Iraqi efforts to escape the strict terms of oil and gas contractual arrangements negotiated since 1992. The intricacies and complexities of the specific language of article 141 of the Iraqi Constitution and article 54 of the KRG oil and gas law (No. 22) were explored. In addition, those provisions of the Iraqi Constitution that serve to safeguard contractual rights and thereby constrain any Iraqi exercise of authority to walk away from commitments were examined. Surveyed as well were the details of apposite provisions of law designed to protect Iraqi oil and gas assets from creditor action. Especially relevant in that respect are the terms of paragraph 22 of Security Council resolution 1483 (as extended and reaffirmed by Security Council resolutions 1859 of December 2008 and 1905 of December 2009) and the language of various national implementing measures, such as U.S. Executive Order 13303 as extended by Presidential Notice of May 20, 2009. Concerning the former, much attention has been given to whether distinctions are to be drawn between the character of the Iraqi oil and gas assets that comprise the object of a creditor's interest, the nature of the proceeding or process utilized to pursue satisfaction against those assets,

the fact that a claim is based on an older rather than a more recent contract, and actions brought against third parties or subcontractors rather than against Iraq or one of its instrumentalities or agencies.

Other than the summary in the preface to situate this study in the larger context of general international law and national law regarding debts from international transactions, it was never a goal to examine more than the narrow and explicit terms and provisions of law specifically applicable to the debtor–creditor question regarding Iraq and the matter of protecting that nation's oil and gas assets so they might be used to facilitate reconstruction and development. The entire issue of creditor claims against Iraqi oil and gas assets might stimulate for some an interest in the various substantive aspects of Iraqi oil and gas law or for others an interest in exploring jurisdictional or procedural difficulties or general recognition and enforcement problems associated with prosecuting claims against Iraqi oil and gas assets. However, the only aspects of international or domestic law concentrated on in this study were those adopted to specifically address the Iraqi situation. The author undertook a preliminary survey of the developing substantive oil and gas law of Iraq in another volume,[1] and jurisdictional, procedural, and general recognition and enforcement-law issues that can arise related to creditor claims against Iraqi oil and gas assets are discussed at length by other scholars writing in other contexts. By focusing on the limited matter of international and domestic law specifically addressing creditor claims against Iraqi oil and gas assets, attention was directed to a topic that otherwise has not been the subject of rigorous analysis and consideration. Chapters 5 through 8 may be tedious in their analysis and consideration, given both their intense concentration on and dissection of the specific words of relevant legal measures and the fact that each chapter examines essentially the same measures with only the targeting of different words to distinguish them. The usefulness of such intensive scrutiny is not likely to be underestimated by those who daily labor in the detail of the law affecting claims against Iraqi oil and gas assets.

Two particular points require reiteration here and, in one sense, they are closely linked. First, it is clear that the insulation of Iraqi oil and

[1] See Rex J. Zedalis, *The Legal Dimensions of Oil and Gas in Iraq: Current Reality and Future Prospects* (Cambridge University Press, 2009).

gas assets provided by the terms of Security Council resolution 1483, through its extension in resolutions 1790 and then 1859, is scheduled to expire at the end of 2010 pursuant to the terms of Security Council resolution 1905. Similarly, according to Presidential Notice of May 20, 2009, the protection provided to those same assets against creditor action in the United States expires in May 2010. Although at some point there likely will be prevailing pressure to remove the protection that benefits Iraqi oil and gas assets, it is reasonable to expect that Iraq's indubitable need for significant revenues, its continuing perilous stability situation, and the downward pressure placed on world oil and gas prices by the current and almost unparalleled international economic crisis all seem to coalesce to postpone that day of reckoning, thereby increasing the probability of further extensions in the future. Until then, the various issues explored herein remain of intense interest to Iraq's creditors and their legal counsel.

Second, irrespective of when the protection currently provided Iraqi oil and gas assets ends, the basic structure and many attributes that accompany it offer considerable lessons for other situations in which resource-rich nations owe debts to foreign creditors. Admittedly, the peculiarities of each situation differ; thus, the nature of the protective regime under consideration must be tailored accordingly. Even in the event that the insulation of Iraqi oil and gas assets from creditor claims were allowed to expire at the end of their present term, the lessons associated with the structure of that regime would endure. Consequently, regardless of whether the forces favoring termination of the extant regime or those that currently militate in favor of its extension prevail when debate resumes, the lessons emerging from the entire experience of protecting Iraq from legal claims related to its oil and gas assets and dealings remain valuable.

Recent developments suggest the existence of encouraging signs regarding Iraq's future. Contending ethnic and sectarian factions have demonstrated the ability to accommodate one another and exhibited the beginnings of tolerance for opposing perspectives. No one knows with certainty whether this will translate into the foundational requisites essential for the success of democratic self-governance. What is known beyond any doubt, however, is that Iraq's oil and gas assets are key to its economic survival, and on that survival hinge the prospects for stability, reintegration as a full member into the world community,

and eventual development as a substantial contributor to the international economy. It cannot be denied that debts legitimately owed by Iraq to foreign creditors must be satisfied through payment or resolved through negotiation or appropriate legal action. At the same time, there is no question that Iraqi oil and gas assets are vital to its national survival and to the spilled blood and spent treasure since 2003 not being rendered in vain. Although the drafters of the relevant international and national measures concerning Iraqi oil and gas assets and dealings structured them in a way that presents various complexities, it is reasonable to expect that decisions about extending their protective aspects will be made with great circumspection and temperance. Only time will tell.

INDEX